PITTING CORROSION

Edited by **Nasr Bensalah**

Pitting Corrosion
Edited by Nasr Bensalah

Published by InTech
Janeza Trdine 9, 51000 Rijeka, Croatia

Publishing Process Manager Jana Sertic

Technical Editor Teodora Smiljanic

Cover Designer InTech Design Team

Additional hard copies can be obtained from orders@intechopen.com

Pitting Corrosion, Edited by Nasr Bensalah
p. cm.
ISBN 978-953-51-0275-5

For EU product safety concerns:
IN TECH d.o.o., Prolaz Marije Krucifikse Kozulić 3, 51000 Rijeka, Croatia,
info@intechopen.com or visit our website at intechopen.com.

Meet the editor

Nasr Bensalah, Ph.D., H.D.R., is Associate Professor of Electrochemistry at University of Gabes, Tunisia, and Research Scientist at Texas A&M University at Qatar, Qatar. He received his B.Sc. degree in Chemistry in 1995, M.Sc. degree in Physical Chemistry in 1997, Ph.D. degree in Electrochemistry in 2002 from Faculty of Sciences of Monastir, Tunisia, and H.D.R. degree in Environmental Electrochemistry in 2007 from Faculty of Sciences of Tunis, Tunisia. Dr. Bensalah held the positions as Senior Lecturer from 1999 to 2003, Assistant Professor from 2003 to 2007, and Associate Professor since 2007 at University of Gabes, Tunisia. He taught courses related to Analytical Chemistry, Environmental Chemistry, and electrochemistry for undergraduate and graduate students.His awards include a Fulbright Scholarship from U.S. Department of State in 2008, AUF Fulbright fellowship in 2006, AECI Spanish government scholarship in 2004. During his academic career, He had been involved in many administrative positions at departmental and university levels. His research interests cover the fields of Electrochemical and Environmental Engineering. His research activities are mainly focused on the electrochemical treatment of water and wastewater, and the monitoring and the control of corrosion, with emphasis on Al and Fe Alloys. Additionally, he conducted some investigations on water and wastewaters treatment by different processes including coagulation and adsorption techniques, disinfection, and advanced oxidation processes. He authored and coauthored about 50 papers in recognized international journals and he supervised more than 10 graduate students.

Contents

Preface

The Pitting corrosion book covers recent research challenges and various accomplishments in the field of corrosion. The book is a proven record of excellent scientific achievements in the field of pitting corrosion.

Pitting corrosion involves local dissolution followed by the formation of cavities on metals surfaces coated with a passive film, when exposed to an aqueous solution containing aggressive anions, such as chloride and sulfate. Pitting is one of the most destructive forms of corrosion, as it causes potential failure of metals and alloys due to perforation/penetration. Pitting corrosion can be quite small on the surface and very large on the subsurface. The process consists of various stages including passive film breakdown, metastable attack, stable growth and pit stifling or death. The breakdown of the passive film is required for pits to initiate. The metastable pitting phase can be considered as a key stage of pitting corrosion, since only pits that continue to grow become stable growing pits. Pitting is influenced by metal composition, electrode potential, temperature, surface condition, and environmental parameters such as aggressive ion concentration, pH, and inhibitor concentration.

Accurate and efficient detection and monitoring of pitting corrosion is very challenging. Pitting failures can occur unexpectedly, and with minimal overall metal loss. The monitoring of pitting corrosion can be further complicated by a distinction between the initiation and propagation phases of pitting processes. Developing and understanding the mechanisms by which pits form and propagate may allow engineers to build more accurate corrosion models in order to predict the lifetime of components and, thus plan when inspection and replacement are needed.

The first chapter addresses the electrochemical characterization of pitting corrosion with an emphasis on the behavior of beryllium using single and poly crystalline scales under different conditions, such as the presence of chloride, using SEM examination techniques. The second chapter discusses the general mechanisms of pit growth during pitting corrosion in general, using Al and Al alloys as examples while focusing on pitting morphology, pit propagation, and hydrogen evolution results.

Pitting corrosion monitoring and detection using acoustic emission is covered in chapter three using detailed discussions of the theoretical and experimental design of such testing techniques with application to stainless steel. Chapter four discusses the use of oscillatory phenomena and non-linear dynamic behavior as a probe to study pitting corrosion of iron in halide-containing sulfuric acid solutions. Chapter five deals with the effect of micro-structural heterogeneity on pitting susceptibility, with an application to induction-quenched steel pipeline using different metallographic examination techniques.

Chapter six discusses a very interesting topic related to associated toxicity and biological risks of pitting corrosion of metallic biomaterials, used as dental and orthopedic implants, with emphasis on biological, sanitary, metallurgical, economical, and social point of view.

In chapter seven, positive application to the pitting phenomena, such as the importance of etch film formed during an alternating current electrograining mechanism, as well as the morphology for pitting of aluminum surfaces to obtain high surface area. The chapter discuses in details the different conditions and parameters that affect the AC electrograining process in relation to the final pitting morphology.

Chapter eight discusses the corrosive effect of chloride ions on metals, with emphasis on pitting corrosion and fatigue of stainless steel in relation to changes in microstructure and the durability of the metal.

Taking into account that corrosion is costly and dangerous phenomenon, it becomes obvious that people engaged in the design and the maintenance of structures and equipment, should have a basic understanding of localized corrosion processes. The Editor hopes that this book will be helpful for researchers in conducting investigations in the field of localized corrosion, as well as for engineers encountering pitting and crevice corrosion, by providing some basic information concerning the causes, prevention, and control of pitting corrosion.

Dr. Nasr Bensalah
Associate Professor,
Faculty of Sciences of Gabes,
Tunisia

1

Electrochemical Characterisation to Study the Pitting Corrosion Behaviour of Beryllium

J. S. Punni*
AWE, Reading
UK

1. Introduction

Beryllium has widespread uses in aerospace industry as it has attractive mechanical properties, a high melting point (1289 °C), a low density (1.85 g/cc), high specific heat capacity and thermal conductivity. It has a hexagonal close packed (hcp) structure and due to its low neutron cross section it is widely used for nuclear applications. To achieve the required mechanical properties beryllium is produced by vacuum hot pressing in the temperature range 1000 to 1100°C, using a high purity and fine grained beryllium powder. To get stress relief, the material is subsequently heat treated at 800°C; this also serves to remove elemental aluminium at grain boundaries in material structure, by converting it to the intermetallic form $AlFeBe_4$. A proper balance between Fe and Al is required to avoid 'hot shortness' due to the presence of elemental aluminium at the grain boundaries.

The principle contaminants within commercial beryllium are oxygen (as beryllium oxide), carbon, silicon, iron, aluminium and magnesium. Silicon, iron and aluminium principally come from the ore, although additional iron may be contributed from billet machining operations. Magnesium arises primarily from reduction of beryllium fluoride using magnesium to produce beryllium metal. Carbon arises principally from casting operations, which use graphite moulds. Oxygen is always present as an oxide film on powder particles. Other elements are also present but generally at very low levels.

Past studies have shown that beryllium is susceptible to pitting corrosion in the presence of chloride, fluoride and sulphate ions (Hill, et al., 1996, 1998; Stonehouse & Weaver, 1965). This is due to the breakdown of passive film at localised sites resulting in sporadic pits. The pitting is an insidious form of corrosion since it can proceed unnoticed and can lead to a catastrophic component failure. During the pitting process most of the metal surface remains passive and acts as the cathodic site for a small anodic area inside the pit which leads to accelerated attack in this location.

Pitting corrosion is a localised form of corrosion by which cavities and holes are produced in the material, which are generally plugged with corrosion products. Corrosion, in general, is an electrochemical process in which electrons are generated and consumed at the corroding metal surface. This process consists of (i) an anodic site at the metal surface where metal is

---

converted into metal ions by losing electrons. In the case of beryllium the anodic reaction is as follows:

$$Be = Be^{2+} + 2e^- \quad (1)$$

Beryllium is a reactive metal having a standard potential of E^o = -1.85V and it lies between aluminium (-1.66V) and magnesium (-2.37V) in the reactive series (Kaye & Laby, 1978)

i. A cathodic site at the metal surface where a complementary reaction to consume excess electrons takes place which could be either

$$O_2 + 2H_2O + 4e^- = 4OH^- \quad (2)$$

At E^o = +1.23V – 0.591 pH at 25°C or

$$2H^+ + 2e^- = H_2 \quad (3)$$

In acidic conditions, at E^o = +0.0V – 0.591 pH at 25°C

ii. A conductor to conduct electrons between the anode and cathode and (iv) an electrolyte to provide a medium for transporting ionic products away from the metal surface.

The stability (potential-pH) diagram for beryllium in water at 25°C, as provided by Pourbaix (Pourbaix, 1966), is represented in Figure 1. This defines the regions of corrosion, passivation and immunity. The problem of pitting corrosion is mainly confined to the passivation region where insoluble beryllium hydroxide is a stable product. At low temperature, in high purity water, beryllium has little or no corrosion problem. However, the performance of beryllium in tap water can be seriously compromised by the presence of small concentrations of chloride and sulphate ions (Stonehouse & Weaver, 1965).

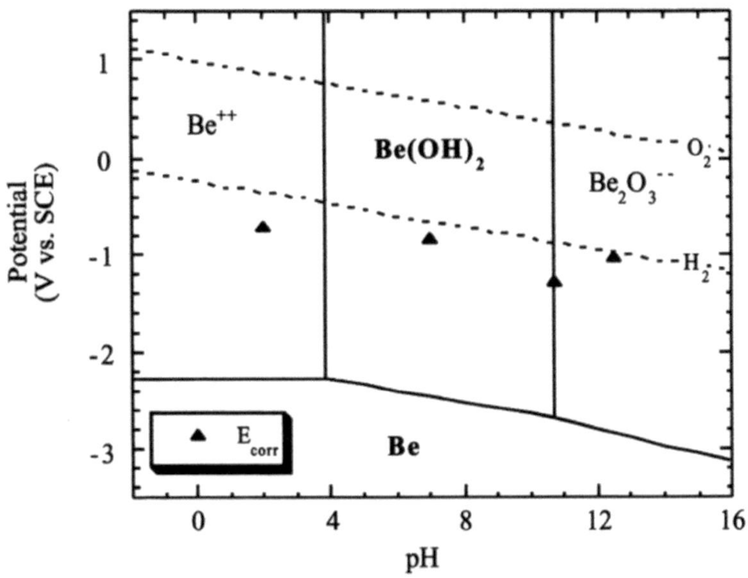

Fig. 1. Potential-pH diagram for the Be-water system at 25°C for the β-$Be(OH)_2$ crystalline hydroxide and assuming a 10^{-6} M concentration of Be^{2+}, after Pourbaix (1966)

2. Review of beryllium corrosion

2.1 Background

Pitting corrosion of beryllium in moist air has been reviewed by several authors (Stonehouse & Weaver, 1965; Miller & Boyd, 1968). Beryllium has adequate corrosion resistance for use in many engineering applications due to the formation of a protective oxide film when exposed to air. The metal will normally remain bright for years upon exposure to the atmosphere. The oxide (BeO) formed on the surface of beryllium is resistant to cracking or spalling because it grows inward and it is in a state of compression (due to its volume being 1.68 times larger than the metal it occupies). However, two prime situations were described by Mueller and Adolphson (1979) under which beryllium was found susceptible to pitting corrosion. The first is when beryllium has carbide particles exposed to moist air which produce active sites for corrosion. This was evidenced by the white corrosion products developed at sites occupied by these inclusions on beryllium under storage conditions. The second is when the surface of beryllium is contaminated with chloride and sulphate ions which may be introduced from human contact, contaminated air, rain water, cleaning solvent or certain packaging materials. A study by Stonehouse and Beaver using humidity cabinet experiments evaluated that machined beryllium (having surface residue deposits) was susceptible to pitting corrosion.

An initial study on the corrosion of beryllium in water focused on reactor environments where beryllium was used as a neutron reflector, and is not available in the general literature. Although beryllium corrosion rates were high in some of the simulated environments, service life for beryllium reflectors exposed to high purity irradiated water was reported to be good. There was no major problem of coupling beryllium with aluminium and stainless steel (Miller & Boyd, 1968). Another study was conducted (Flitton et al., 2002) as part of nuclear waste disposal, to understand beryllium corrosion as a result of being placed underground. Beryllium (99% pure) coupons were buried 4 and 10 feet deep underground. The general loss of material was measured at the rate 2 μm and 7 μm per year respectively, and the maximum pit depth at 4 feet underground depth was recorded as 153 μm. In some other instances where beryllium components were stored in uncontrolled moist environments, corrosion pits as deep as 250 μm were detected.

It has generally been recommended (Mueller & Adolphson, 1979; Stonehouse & Weaver, 1965) that for beryllium, the use of tap water should be minimised, metal surfaces should be as clean as possible, and excessive machine damage should be avoided. The storage of finished beryllium components should be maintained in polythene bags with a desiccant to maintain a dry atmosphere and in humidity controlled room. Where corrosion conditions are expected in service, the use of chromate coating and anodic films can be beneficial.

2.2 Effect of pH and chloride ion concentration

Electrochemical techniques were widely used in the past (Hill, et al., 1996, 1998; Punni & Cox, 2010; Lillard, 2000) to characterise pitting corrosion behaviour of beryllium in aggressive aqueous environments. In aqueous salt solutions beryllium develops a passive hydrated oxide film, either $Be(OH)_2$ or its hydrate ($BeO\ (H_2O)x$) (Vaidya et al. 1999) Electrochemical impedance spectroscopy (EIS) experiments had shown that the oxide growth rate on beryllium was 6.4 angstroms/V over the range of 0–4V. This barrier layer is quickly disrupted at localised sites in the presence of aggressive ions in the environment (Hill et al., 1998).

The passivity and pitting of commercially produced beryllium has been studied extensively in chloride solutions. The key points for a typical cyclic polarisation curve from a recent polarisation study (Punni & Cox, 2010) for a commercial beryllium are displayed in Figure 2. It was observed that the anodic polarisation was characterised by a region of passivity followed by a logarithmic increase in the current density, which corresponded to the onset of pitting corrosion (E_{pit}).

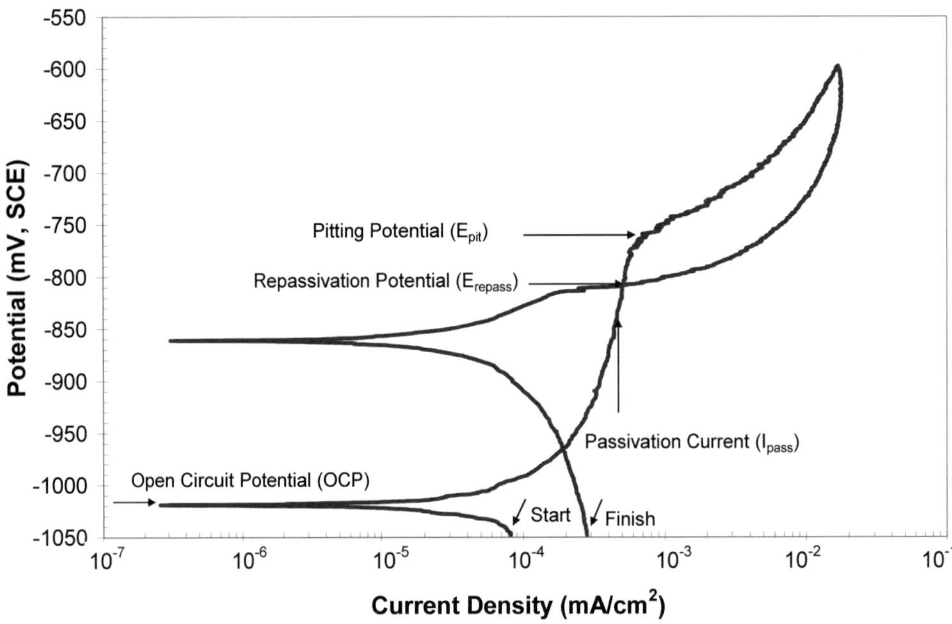

Fig. 2. Typical cyclic polarisation curve for beryllium metal in 0.0001M KCl solution, sweep rate of 20mV/min, after Punni & Cox (2010)

A previous electrochemical study (Hill et al., 1996, 1998) in sodium chloride (NaCl) solution of varying chloride ion concentrations (from 0.0001 to 1M) showed that on increasing the chloride concentration, both pitting potential and repassivation potential decreased progressively and at 1M NaCl, the pitting commenced at open circuit conditions. A logarithmic relationship between pitting potential (in mV) and chloride concentration (in morality) was presented as given below.

$$E_{pit} = -67 \log [Cl^{-1}] - 1010 \quad (4)$$

The effect of pH on pitting corrosion behaviour of beryllium was studied by Hill and co-workers (1998). Beryllium was found to be passive in the pH range from 2 to 12.5 but it was susceptible to general attack below a pH of 2. The surface of samples exposed to pH 1 solutions were characterised by thick, black deposits over the entire surface. There was an increase in pitting potential with increase in pH. The total change in pitting potential over the pH range 2 to 12.5 was ~35 mV.

According to another investigation (Friedman and Hanafee, 2000) the change in pitting potential was rather greater ~90 mV over the same sort of pH range. This investigation also found that chloride concentration in the range 0.001 to 0.1M NaCl did not have a significant effect on corrosion of beryllium. This conclusion was based on the observation that there was not much difference in passive range. It was stated that the size of the passive range, or the ability of a solution to break the oxide layer on the surface of the beryllium, appeared to be a function of pH, and not of chloride concentration. This result on chloride effect is not in agreement with the findings of Hill and co-workers (1996) and a recent investigation by Punni and Cox (2010).

A minimum in both the passive and corrosion current densities was observed in the pH range of 4 to 11 (Hill et al., 1996, 1998). The relationship of both passive current density and corrosion current density with solution pH is represented in Figure 3. This agrees well with the Pourbaix diagram (Pourbaix, 1966) for the beryllium water system at 25°C, as represented in Figure 1. This shows that between pH 4 and 11, the formation of insoluble species $Be(OH)_2$ is thermodynamically favourable.

2.3 Crystallographic orientation effect

The crystallographic orientation effect of commercial beryllium was described by Lillard (2000) and Friedman and Hanafee (2000). Pitting experiments indicated that once pit initiation has started, beryllium corrodes preferentially in certain orientations. Potentiodynamic polarisation studies for beryllium (0001), (1010) and (1120) surfaces were carried out in deaerated 0.01 M NaCl, as represented in Figure 4. The pitting potential (E_{pit}) was found to decrease with crystal orientation in the order (0001) > (1010) > (1120). Repassivation potentials (R_{rp}) appeared to follow the same crystallographic trends as E_{pit}. The potential region defined by E_{pit} - E_{rp} was large for (0001) and (1010) surfaces (>100mv) and appeared to be much smaller for the (1120) surface. There is a relationship between E_{pit} and pit density on the surface such that for a higher E_{pit} at a surface there is a smaller population density of pits and vice versa. The expression E_{pit} - E_{rp} represents the extent of pit growth. This implies that a surface orientation with large E_{pit} and with large (E_{pit} - E_{rp}) would experience fewer, but deeper, pits.

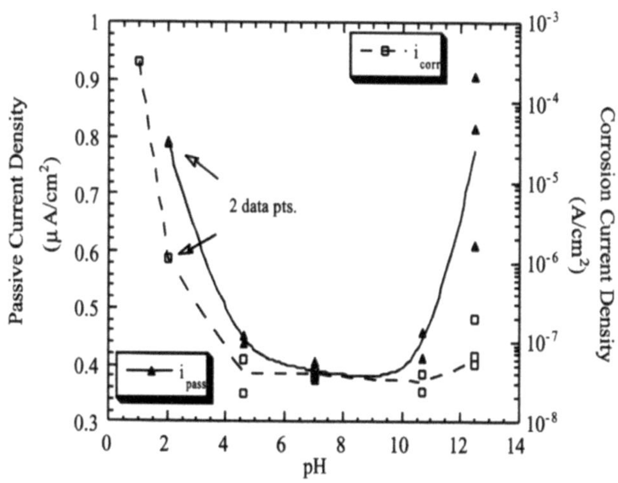

Fig. 3. Passive current density, i_{pass}, and corrosion current density, i_{corr}, as a function of solution pH. A minimum in i_{pass} and i_{corr} exists between a pH of 4 and 11, after Hill et al. (1998)

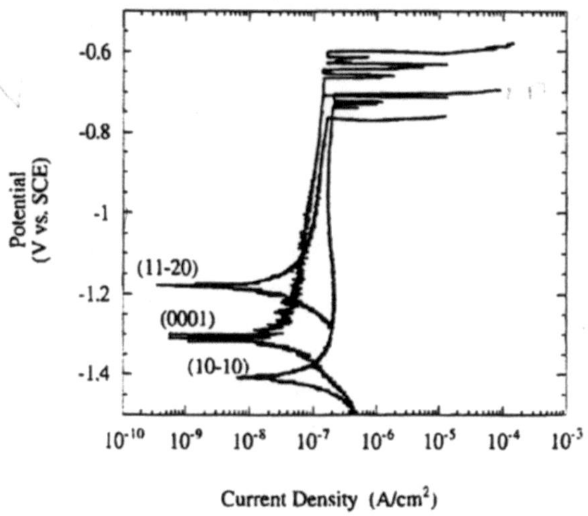

Fig. 4. Typical polarisation curves for (0001), (1010) and (1120) single crystal Be surfaces in deaerated 0.01 M NaCl solution, after Lillard (2000)

Cyclic polarisation studies on beryllium single crystals in 0.01M NaCl solution with a pH of 7, were also conducted by Friedman and Hanafee (2000). The size of the passive range differed in all three planes, but the basal plane differed most compared to the two prism planes. Interestingly, E_{rp} was identical for all three crystal orientations; -864mV. The difference in the results from the two investigations may be explained by variations in electrochemical conditions.

Orientation imaging microscopy maps of polycrystalline beryllium indicated no correlation between pit initiation sites and crystallographic orientation of specific grains. Therefore, altering the bulk texture of polycrystalline beryllium will not affect the pitting corrosion resistance of beryllium, although it may affect pit propagation (Lillard, 2000).

2.4 Morphology of electrochemical corrosion pits

The morphology of corrosion pits in polycrystalline commercial beryllium on electrochemical polarisation in 0.01M KCl solution was described by Hill et al. (1998). It was observed that the corrosion pits in beryllium have the same size and shape as the beryllium grain morphology. The pits were not hemispherical and parallel plates of unattacked beryllium were found inside the pits.

In the case of the single crystal study reported by Lillard (2000) for (1010) and (1120) surfaces, the pit interiors were characterised by crystallographically oriented parallel plates of unattacked beryllium. At greater depth these beryllium lamellae were transformed to small 'wires' or fibres of uniform diameter. For the beryllium (1010) surface, the orientation of the fibres was parallel to the (0001) direction and normal to the (1010) surface. For the beryllium surface (1120), the lamellae were parallel to the (0001) direction and normal to the (1120) surface. In comparison to the (1010) and (1120) surfaces, the interior of corrosion pits formed in the (0001) surface lacked any crystallographic orientation; however, the propagation was often in the (1010) and (1120) directions. As in the case of polycrystalline material, the small fibres of unattacked beryllium were also observed in the interior in the (0001) surface but the orientation of these fibres was normal to (0001) surface. Similar differences in pit morphologies between the basal plane (0001) and prism planes (1010) and (1120) were also observed by Friedman and Hanafee (2000).

2.5 Performance of pitted beryllium under stress

In a previous investigation (Vaidya et al., 1998) pitted beryllium samples were tested under 4-point bending stress. Beryllium samples were exposed to 0.01, 0.1 and 1 M NaCl solutions for 168 hours. The resultant bending stress results (displacement to failure plots) are shown in Figure 5. The material exhibited a typical elastoplastic response, with a small amount of strain hardening. Displacement to failure and failure strength of the samples were affected by exposure to the NaCl solutions, however, the displacement to failure was much lower in 0.1 M and 0.01 M solutions than in a 1 M solution.

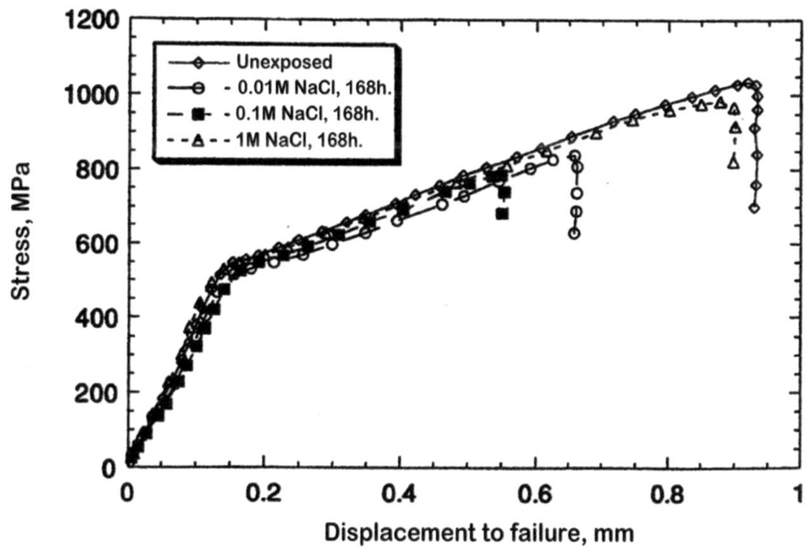

Fig. 5. Bending stress vs displacement to failure plots for as received and corroded beryllium samples after Vaidya et al. (1998)

This was attributed to difference in the sizes and distribution of the pits formed on the surface of the samples. The fewer but deeper pits which formed on the samples exposed to 0.01 and 0.1 M NaCl solutions act as stress concentration sites and degrade the mechanical properties of the material to a greater extent than those exposed to 1M NaCl solutions.

2.6 Mechanism of pitting corrosion

The pitting corrosion process for beryllium is considered to be common with other metals and alloys (aluminium, titanium and stainless steel) which develop a passive oxide film. Following the development of the passive oxide film there are four processes (Szklarska-Smialowska, 1999) which lead to the onset of corrosion pits as; (1) Interactions occurring at the surface of the passive film; (2) interactions occurring within the passive film; (3)

formation of metastable pits which soon repassivate; and (4) stable pit growth, above certain potential termed the critical pitting potential.

The first two stages are concerned with the breakdown of film by the interaction of chloride ions, with very little being known about the interaction mechanism. This is certainly dependent upon the composition and structure of the oxide film and the underlying material. The structural characteristics of the oxide are dependent on the material composition and hence the presence and distribution of micro-defects (vacancies, voids, etc.) as well as macro-defects (inclusions, second phase particles and their size and shape), crystal structure and the degree of noncrytalinity of the oxide. It also depends on the electrolyte, its composition, pH value and temperature (Szklarska-Smialowska, 1999).

Metastable pits form sometimes just before the pitting potential is reached. When these occur first the current increases as the pit nucleate and grow, then decreasing after a short delay. It was found that metastable pits are covered by a remnant hard oxide film. It is well established that metastable pits sometimes form several hundred milli volts below the pitting potential and during the induction time for stable pit formation. There are a number of studies for metastable pits on steel and aluminium (Blanc & Mankowski, 1997; Pride et al., 1994) but few on beryllium (Lillard, 2000).

According to a previously suggested mechanism (West, 1970) a filmed metal may begin to dissolve at sensitive points. These sensitive points may be crystallographic defects, cavities or scratches in metal, or a rift in surface film due to high internal stresses during oxide growth. Some of these sensitive points are associated with the localised chemical deficiencies in the film where there are underlying inclusions. As the dissolution proceeds at localised points where the film is impaired, the remaining intact surface film acts as a cathodic area. At this stage, two other important processes are taking place inside the pore (a shallow pit). Firstly the attempt by the film to repair itself at the pore consumes hydroxyl ions so the acidity within the pore is increased. Secondly the pore being an anodic site, causes various aggressive ions such as Cl^{-1} to accumulate within the pore. Being highly deformable, chloride ions are surface active, so they displace water molecules from the double layer. The film in the pore therefore is prevented from repairing itself. At the same time, depletion of aggressive ions from the immediate vicinity of the pore makes it more difficult for the anodic reaction to spread sideways. This action together with a large surface area acting as cathode makes an accelerated dissolution at the localised area of pore, thus promoting it to become a narrow deep pit.

3. Investigation on the effect of inclusions on pitting corrosion behaviour of beryllium

3.1 Introduction

Mueller & Adolphson concluded that pitting corrosion is associated with beryllium carbide particles. Carbon arises as one of the primary impurities in historic grades of beryllium to 0.2 wt% or more, which results in formation of second-phase particles (Be_2C). In the presence of moisture these particles quickly react to form BeO and methane.

$$Be_2C + 2\,H_2O \rightarrow 2\,BeO + CH_4 \quad (5)$$

It may be noted that once carbide particles have been consumed no further corrosion should be observed. However, in the presence of corrosive ions in moist environments the above corrosion reaction may disrupt the continuity of the passivating film and provide a crevice site for further reaction.

Earlier studies (Gulbrandsen & Johansen, 1994; Venugopal et al., 2000) have indicated that corrosion pits in beryllium may be associated with sites previously occupied by inclusions. Beryllium components for structural applications are generally produced by a powder metallurgy route and contain a number of impurities as a result of the powder production process, from which the pressed component is manufactured, and as a result of the powder pressing process itself. The major impurities found in such structural material include, beryllium oxide (> 0.5 wt. %), carbides (carbon up to 0.015 wt. %), silicon (~ 0.025 wt. %), and the intermetallic compounds.

The purpose of the study at AWE was to understand the role these inclusions play in the initiation of localised corrosion attack. To accomplish this, the electrochemical characteristics of beryllium material containing differing impurity contents and distribution were assessed and the resulting localised corrosion sites evaluated. Earlier results of this investigation were published in a previous paper (Punni & Cox, 2010) and a brief account of this work with updated information is described here.

3.2 Experimental procedure

3.2.1 Material

Samples measuring 12.5 mm in diameter and 2 mm thick were machined from three different beryllium pressings, i.e. a small vacuum hot pressed bar produced from Brush Wellman S65 specification power (S65-Bar), a large vacuum hot pressed billet also produced from S65 specification power (S65-Billet), and a small vacuum hot pressed bar produced from Kawecki Berylco P10 specification powder (P10-Bar). All pressings had been consolidated using graphite dies under a pressure of ~7 MN/m^2 and at a temperature of 1000 to 1100 °C, followed by a heat treatment at 800 °C. The metallic impurity content for each pressing was determined by Emission Spectroscopy via Direct Coupled Plasma, BeO by Inert Gas Fusion (LECO method), and carbon by combustion (LECO method). The chemical composition of these grades is given in Table 1.

3.2.2 SEM examination and pit initiation treatment

To understand the pit initiation sites it is important to know the microstructure for all the three beryllium grades. For this task, samples were progressively polished down to 0.25 μm diamond finish. Polished samples were examined in a Hitachi SEM with the impurity inclusions being analysed using the associated EDS system (Oxford Instruments). Secondary electron (SE) images and EDS spectra showing elemental analysis were recorded.

For pit initiation treatment the same polished samples were subjected to an electrochemical polarisation treatment to initiate tiny corrosion pits at the polished surface. This was done in 0.001 M KCl solution at a fast sweep rate of 600 mV/min. Polarisation started at -1100 mV SCE and ended at -500 mV SCE and the time during which the sample stayed in the pitting range (-800mV SCE to -500mV SCE) was ~30 seconds; referred to as the exposure time for pit initiation treatment. After the pit initiation treatment the samples were rinsed in distilled water and wiped with ethanol soaked tissues. Samples were then examined again in the Hitachi SEM to locate any pit initiation sites and to perform elemental analysis on those sites.

Element/Compound	S65-Bar	S65-Billet	P10-Bar	Uncertainty of Measurements
Be	99.3	99.23	98.46	Balance
BeO	0.56	0.6	1.18	± 0.04
C	0.03	0.015	0.08	± 0.009
Si	0.025	0.028	0.028	±0.0024
Al	0.02	0.033	0.045	± 0.0020
Fe	0.056	0.075	0.133	±0.0032
Mg	0.0005	<0.0005	0.013	± 0.0023
U	0.004	<0.0030	0.027	± 0.0043
Co	0.001	0.0007	0.0012	± 0.0003
Cr	0.0025	0.0025	0.0135	± 0.0005
Cu	0.0025	0.0065	0.007	± 0.00016
Ti	0.0045	0.011	0.011	± 0.0003

Table 1. The compositions (wt %) of three different beryllium specifications studied in this investigation

3.2.3 Polarisation technique

This technique is described in greater detail in a previous investigation (Punni & Cox, 2010). Samples for polarisation tests were ground down to a 600 grit silicon carbide paper finish. A custom designed glass electrochemical cell, equipped with a platinum counter electrode, noise electrode and a saturated calomel reference electrode (SCE) through a Luggin probe, was used in this study.

The potentiodynamic polarisation tests were carried out at room temperature (20 ± 2 °C) at a sweep rate of 20 mV/min using a Gill AC Potentiostat. Some polarisation tests were carried out at a sweep rate of 10 mV/min. Tests were performed in a deaerated 600 ml potassium chloride (KCl) solution of 0.0001, 0.001, 0.01 or 0.1 M chloride concentrations with pH adjusted to 7, by adding a few drops of dilute potassium hydroxide solution. To ensure dynamic equilibrium, samples were left at their open circuit potential for 1 hour before each polarisation test was run. Some electrochemically polarised samples were also examined in a Hitachi SEM model S3400N to characterise the extent and nature of the corrosion pits.

3.3 Results and discussion

3.3.1 Microstructures of the three beryllium grades

SEM examination revealed that inclusions, varying in size and composition, were situated at the grain boundaries, as summarised in Table 2.

Coarse Inclusions: All beryllium grades showed some coarse inclusions, 5 to 18 μm in diameter. The population and size of these coarse inclusions were greater in P10-Bar than in the other two grades. These coarse inclusions along with the fine inclusions (< 5 μm in size) can be seen in SEM images shown in Figures 6 (a to c). Most of the coarse inclusions appeared to be partially broken or dissolved by the abrading and polishing operations. The EDS analysis identified the coarse inclusions to be mainly elemental silicon, and mixed beryllium carbides containing Al, Si, Fe and occasionally other elements such as Ti, Mg, Cr and U (Table 2).

Inclusion Type	S65-Bar	S65-Billet	P10-Bar
Mixed beryllium carbides: containing Si, Al, and Fe, sometimes had association of Ti, Cr, Mg or U	Mainly detected as coarse type (5-10 μm in size)	Mainly detected as coarse (5-12 μm in size)	Mainly detected as coarse type (8-15 μm in size), relatively high population than other two grades
Elemental Silicon:	Detected as coarse type (5-10 μm in size)	Detected as coarse type (5-10 μm in size)	Coarse particle size 5 to 18 μm, relatively high population than other two grades
Beryllium carbides: (possibly Be_2C)	Detected as a fine type (~2 to 4 μm)	Detected as a fine type (~2 to 4 μm)	Detected as a fine type (~2 to 4 μm)
Intermetallic, Al/Fe/Be: possibly $AlFeBe_4$	Fine type (1 to 2 μm size)	Fine type (2 to 4 μm size), relatively larger size than other two grades	Fine type (1 to 2 μm size)
Intermetallic, Si/Ti and Si/Al (with possible association of Be)	Mostly detected as a fine type (~3 to 5 μm)	Mostly detected as a fine type (~3 to 5 μm)	Mostly detected as a fine type (~3 to 5 μm)
Beryllium Oxide: (BeO)	Fine dispersion (0.2 to 0.5 μm in size) but more densely populated than other two grades	Relatively coarser dispersion (0.4 to 1 μm in size)	Relatively coarser dispersion (0.4 to 1 μm in size)

Table 2. Composition and particle size for each type of inclusion in the three beryllium specifications

Fine Inclusions: The fine inclusions (0.25 to 5 μm in size) can be seen in Figures 6 to 8. These inclusions were examined using EDS analysis, as shown in Figures 7 and 8 and recorded in Table 2. Broadly speaking the fine inclusions at grain boundaries can be classified as finely dispersed oxide and intermetallic phases. EDS identified the finely dispersed oxide to contain the elements Be and O and is therefore probably the hexagonal compound BeO. These oxide inclusions were relatively finer in size and more numerous in the S65-Bar than in the other two grades. The intermetallic phases Al/Fe/Be and occasionally Si/Ti or Si/Al

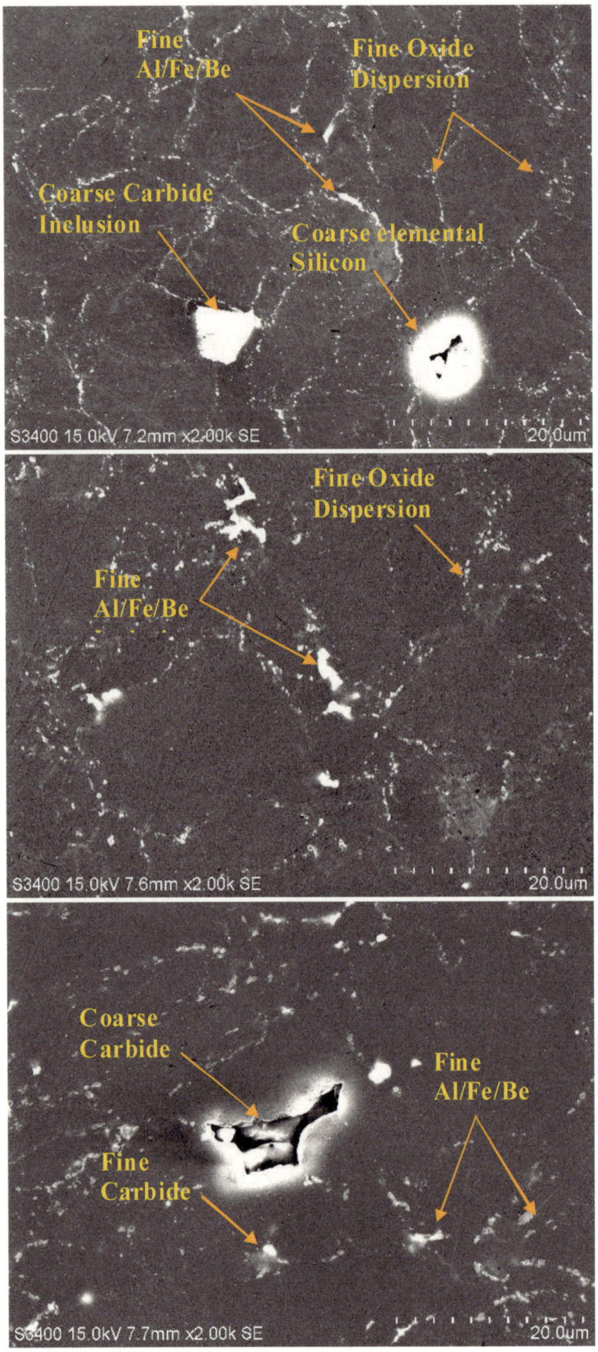

(a) S65-Bar

(b) S65-Billet

(c) P10-Bar

Fig. 6. SEM Images of polished beryllium surfaces from thee beryllium grades

with or without association of Be, were also detected. EDS analysis of the Al/Fe/Be inclusions, showed that the atomic percentage of Fe and Al to be equal, however the Be was unable to be quantified with this technique. From previous findings it is anticipated this intermetallic will be either $AlFeBe_4$ (Carrabine, 1963) or $AlFeBe_5$ (Rooksby, 1962). These inclusions were relatively bigger in size and more numerous in the S65-Billet material compared to the other two grades. Small sized (~3 µm) particles containing only Be and C which are most probably the compound Be_2C were identified in all the three grades. In general, the shapes of the inclusions varied; the Al/Fe/Be inclusions were either L-shaped or elongated while all other inclusions were either round or in some cases elongated.

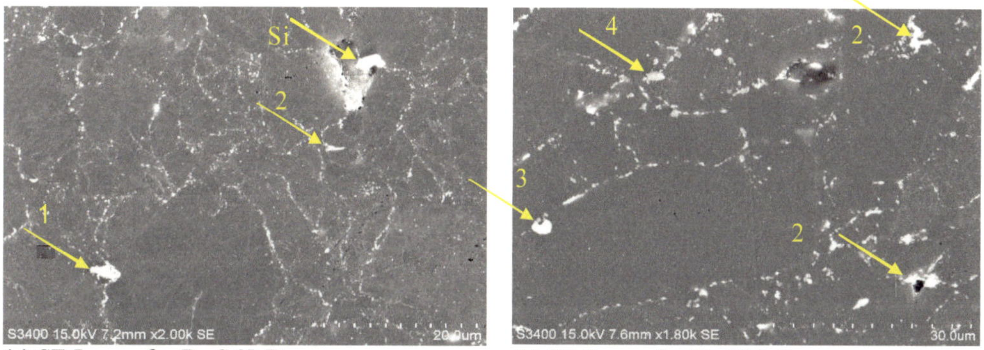

(a) SE Image for Be S65-Bar

(b) SE Image for Be P10-Bar

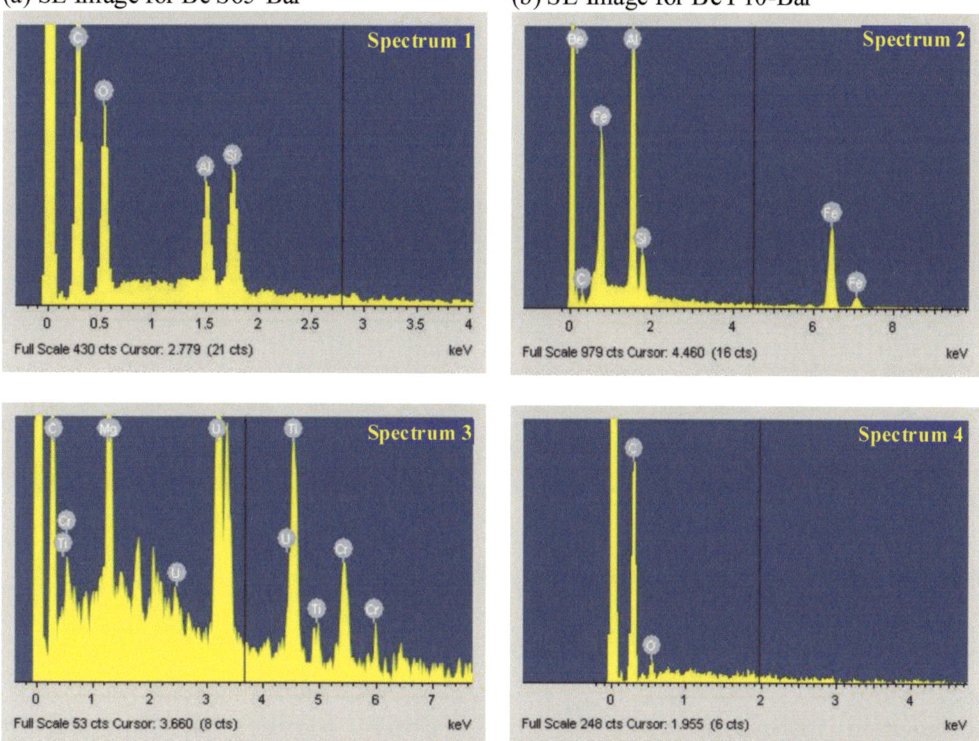

Fig. 7. EDS Spectra from inclusions in beryllium (a) S65-Bar and (b) P10-Bar (a number on an inclusion relates it to the correspondingly numbered typical EDS spectrum)

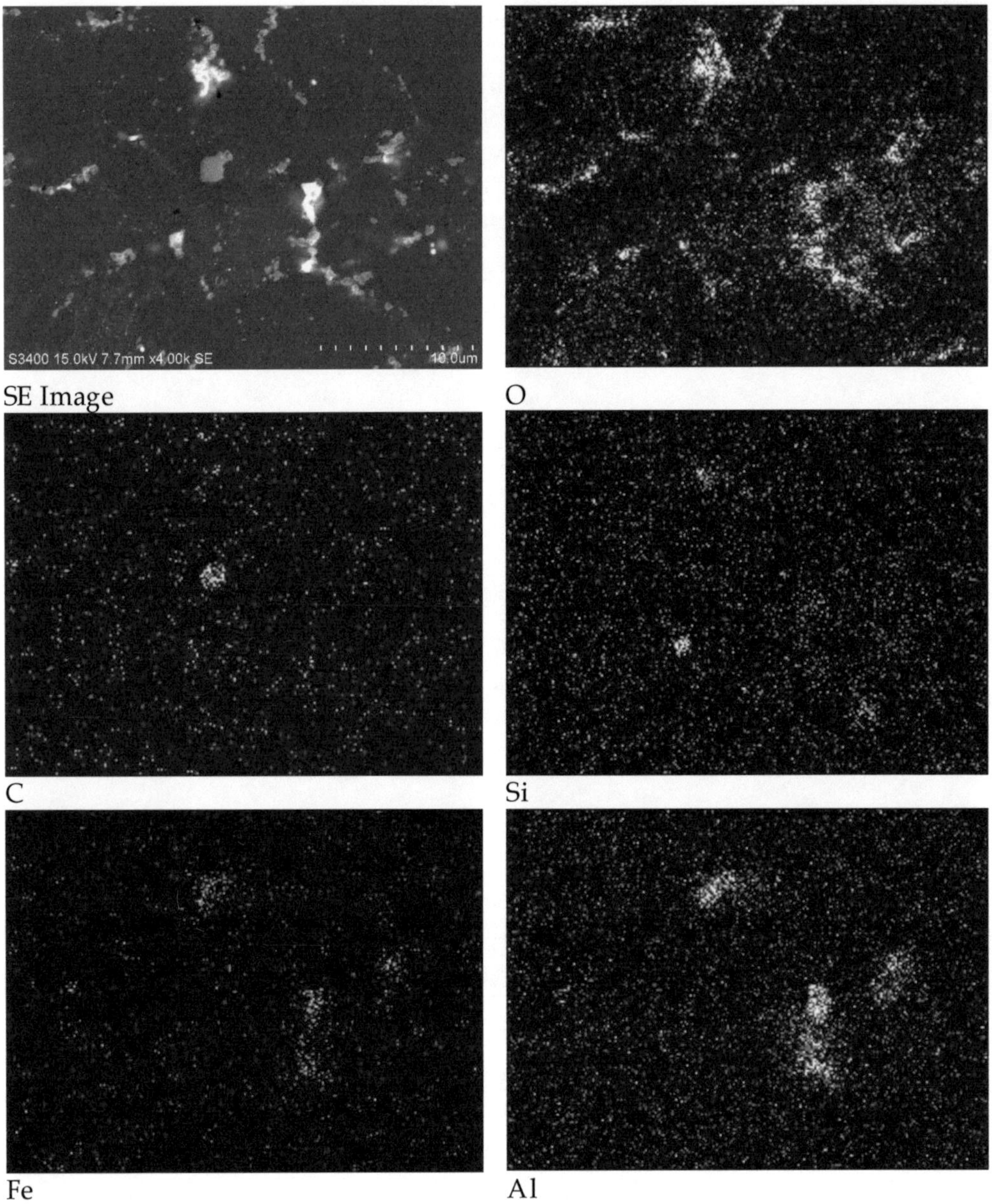

Fig. 8. Elemental images of inclusions in beryllium P10-Bar

3.3.2 Pit initiation sites and pit morphology

Surface morphologies before and after pit initiation polarisation treatment for grade S65-Bar are shown in Figures 9 (a and b). The initiation sites for corrosion pits before and after polarisation are indicated by arrows. From these figures it is clear that corrosion pits nearly always initiated at inclusions in the grain boundaries. In very rare cases the corrosion pits had initiated without any sign of inclusions. As detected by EDS the majority of the pits shown in Figure 9 (b) had initiated at the Al/Fe/Be inclusions and one had initiated at elemental silicon. There was some evidence of pits initiating at mixed beryllium carbide inclusions, and in such cases the inclusions themselves had corroded. This agrees with the pitting mechanism proposed by Mueller and Adolphson that pits initiate at beryllium carbide inclusions. In the case of intermetallic Al/Fe/Be or elemental silicon inclusions, pits had initiated at the interface between the inclusion and the surrounding metal (inclusions being intact at the edge of the pit as seen in Figure 9), indicating that these inclusions are nobler than the beryllium matrix. Work by Vaidya and co-workers (Vaidya et al., 1999) has suggested that BeO particles can also provide pit initiation sites due to their inhomogeneity with the metal matrix. Although a few isolated corrosion pits had initiated without any sign of Al/Fe/Be, elemental silicon or carbide inclusions, there was no conclusive evidence to support the above suggestion that BeO can provide pit initiation sites. Furthermore, no pits were detected at any other type of intermetallic inclusions (i.e. Si/Ti or Si/Al intermetallic particles).

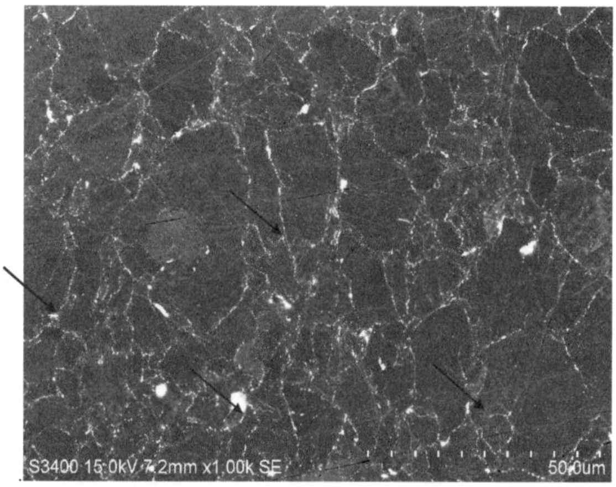

(a) Polished surface of beryllium S65-Bar

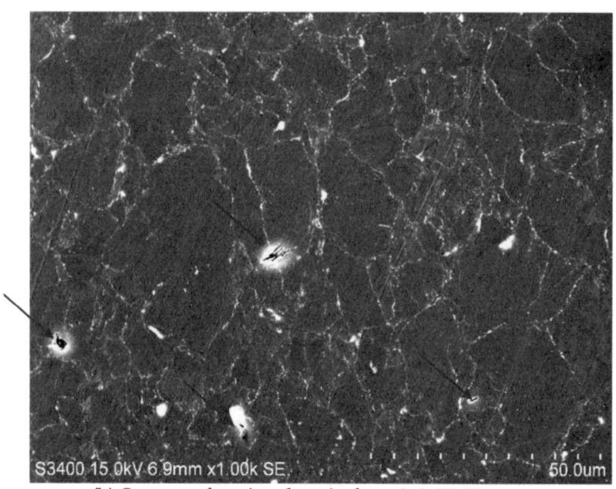

(b) Same surface (as above) after pit initiation

Fig. 9. Comparison of beryllium S65-Bar surfaces before and after pit initiation treatment for 30s exposure in 0.001M KCl

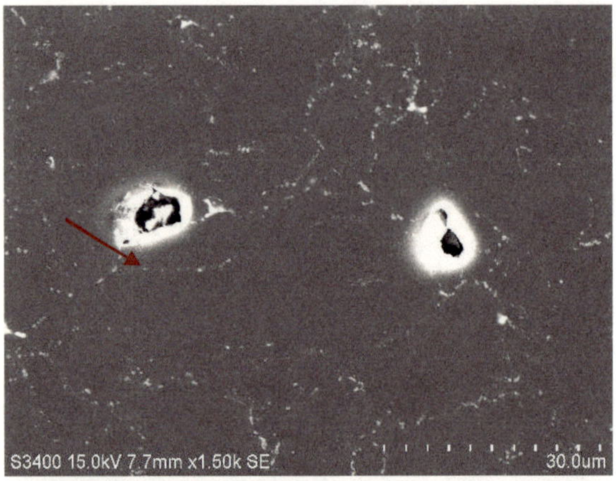

(a) Polished Be surface

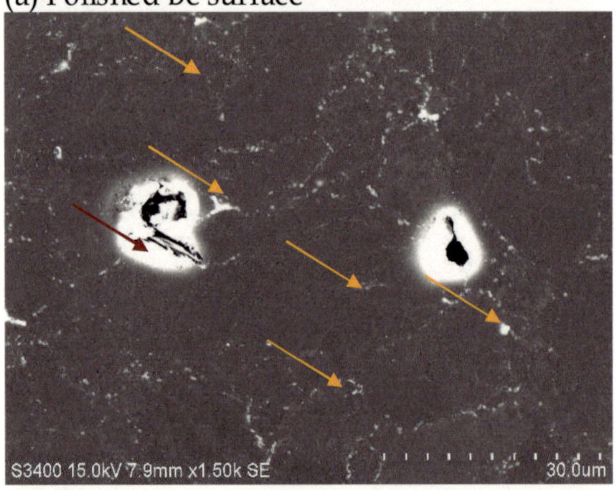

(b) After 1st pit initiation

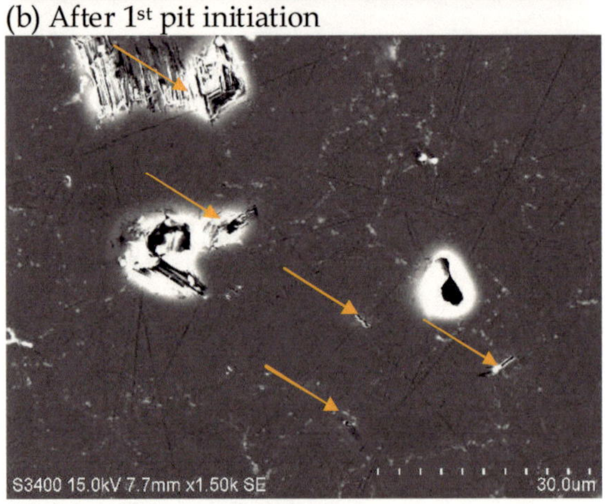

(c) After 2nd pit initiation

Fig. 10. Comparison of beryllium S65-Billet surfaces before and after pit initiation treatment for 30s exposure in 0.001M KCl

(a) Typical topography of an electrochemically produced pit after 30 s exposure in 0.001M KCl

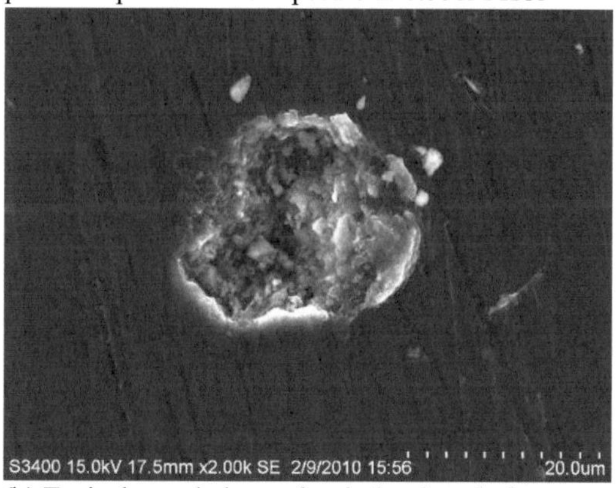

(b) Typical morphology of a pit developed after ageing for 120 days at 75°C in a relative humidity of 16%.

Fig. 11. Typical morphology of corrosion pit produced by electrochemical polarisation and its comparison with that developed during ageing

In the case of S65-Billet, the samples were subjected to two pit initiation treatments, the results of which are shown in Figures 10 (a to c). The results of the first initiation treatment are shown in Figures 10 (a) and (b) where the corrosion pit locations are marked by red arrows. The result of the second initiation treatment is shown in Figures 10 (b) and (c) where the new additional pit locations are marked by orange coloured arrows. It was found that the majority of existing corrosion pits did not grow any further; however, some isolated pits had grown up to 50% of their original size. Apart from the growth of some of the existing corrosion pits, many new pits had initiated, varying in size from 1 to 30 μm. The majority of these new pits had initiated at Al/Fe/Be inclusions as was the case after the first pit initiation treatment.

Another investigation is underway at AWE to study growth kinetics of the corrosion pits on beryllium grade S65-Bar. In this study the corrosion pits were initiated using electrochemical polarisation and subsequently the samples were aged at 55°C, 65°C and 75°C in ovens at 12% to 16% relative humidity. This work is out of the scope for this chapter;

however, some relevant features of corrosion pit formation are described here. This work provided a good comparison between morphologies of electrochemically produced pits with those produced by humidity controlled experiments as shown in Figure 11 (a and b)). Pits produced by electrochemical polarisation showed pit walls and parallel plates of unattacked beryllium lamella left behind after pit propagation, similar to the one observed by Hill et al. (1996). On the other hand pits produced by ageing were covered with corrosion products containing oxygen (main element), silicon, aluminium and trace amount of chlorine. Examination of aged surfaces revealed there was no further growth of large (up to 50 μm) electrochemically produced corrosion pits. In addition to these large pits, there were very small corrosion or etch pits (<3 μm) which had initiated at the intermetallic Al/Fe/Be or elemental silicon inclusions and these grew further on ageing, as illustrated in Figures 12. Figure 13 shows morphologies of some other corrosion pits formed at the inclusion particles, this confirms that even under the controlled humidity conditions, Al/Fe/Be and elemental silicon inclusions are the preferential sites for pit initiation.

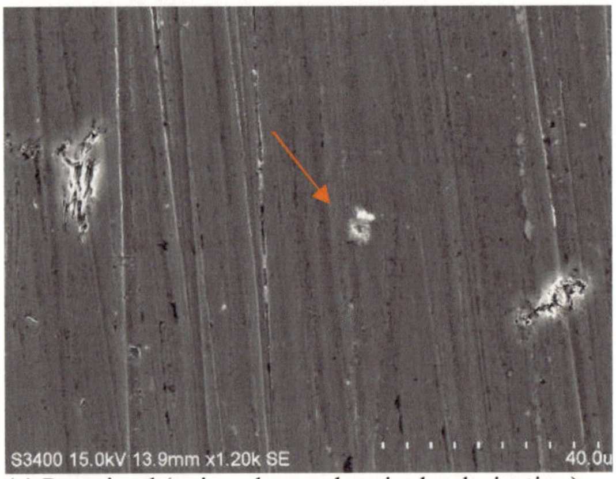

(a) Pre-pitted (using electrochemical polarisation) surface before ageing

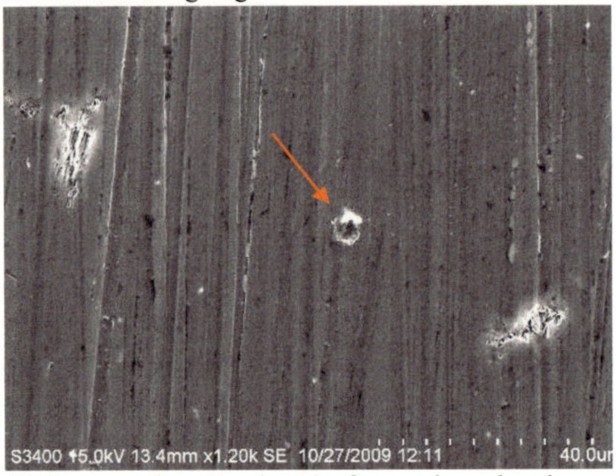

(b) Same surface (as above) after ageing, showing a pit growth at an inclusion

Fig. 12. Secondary electron images showing pit growth (at a site marked by arrows) on ageing at 55°C for 60 days in an environment with RH 16%

There is a similar evidence of pit initiation at the intermetallic inclusions in other passive metals i.e. sigma phase in duplex steels (Mathiesen & Hansen, 2010) and Al_3Fe, Al/Cu/Mg, Al/Cu/Fe and Al_2Cu in aluminium (Szklarska-Smialowska, 1999). All these inclusions are suggested to act as the cathode to initiate the corrosion reaction in the adjacent matrix. There could be other Al/Mg/Mn, Al/Mn/Cr and Al/Mn/Si inclusions in aluminium which are anodic to matrix and are harmless but pits may nucleate at these sites due to dissolution of these particles themselves (Szklarska-Smialowska, 1999). The geometry of inclusions may also affect the pitting behaviour. The difference in the pitting behaviour of 6056 and 2024 aluminium alloys was explained by the difference in behaviour of coarse particles in these alloys (Blanc & Mankowski, 1997).

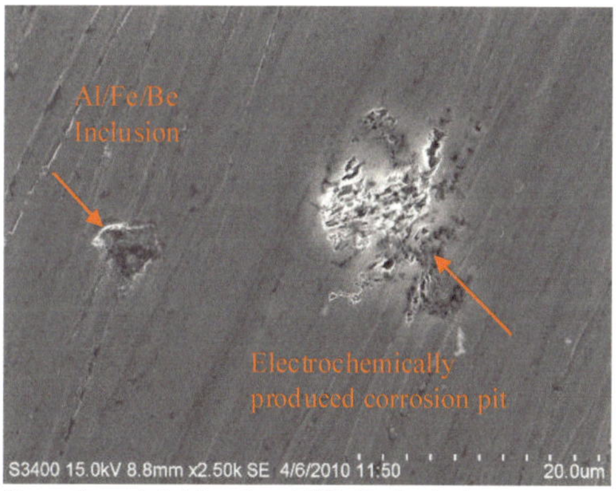

Corrosion pit initiated at Al/Fe/Be inclusion

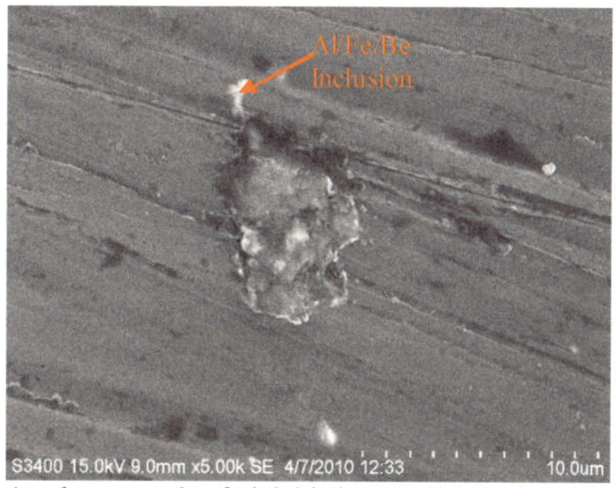

Another example of pit initiation at Al/Fe/Be inclusion

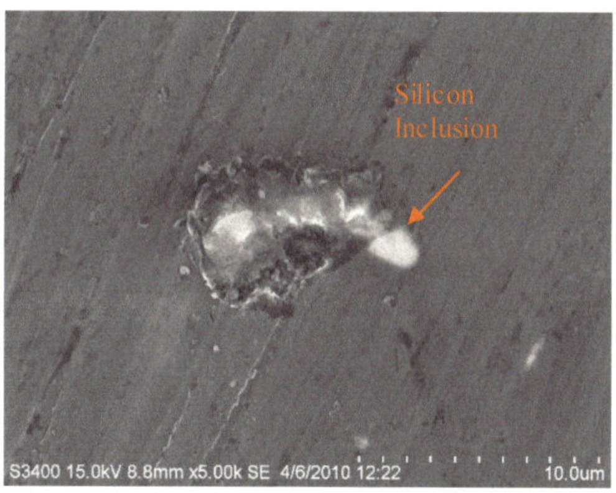

Corrosion pit initiated at elemental silicon inclusion

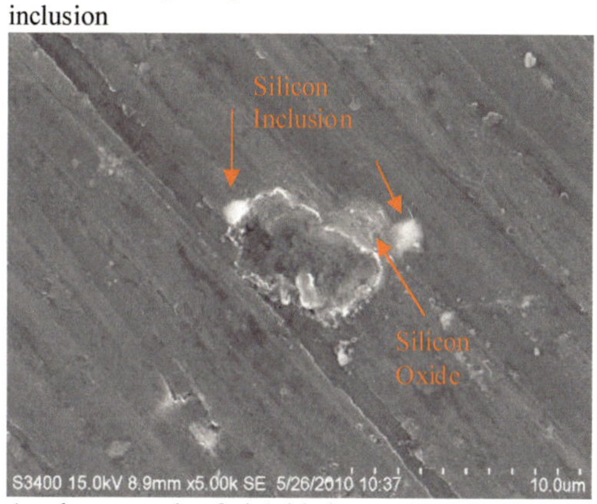

Another example of pit initiation at elemental silicon inclusion

Fig. 13. Morphologies of typical corrosion pits initiated at Al/Fe/Be or silicon inclusions after ageing of pre-pitted surfaces at 55°C for 120 days in an environment with RH 16%

As discussed in section 2.6, the mechanism of film breakdown by inclusions is not clear at this stage, however, it is anticipated that composition or structure of passive film is altered at the site of underlying inclusions. The effected oxide at these localised sites could be more reactive to the chloride ions and result in film breakdown there (Blanc & Mankowski, 1997;

West 1970). It is also evident from this investigation that certain inclusions cause galvanic action with the matrix to initiate pitting, as the corrosion pit sites had the presence of intermetallic Al/Fe/Be and elemental silicon inclusions (Figure 13) suggesting that these act as cathode to promote corrosion of the surrounding metal . Other inclusions like carbides or any less noble inclusions may also result in pit initiation by dissolving themselves preferentially to the matrix.

3.3.3 Electrochemical polarisation of the three beryllium grades

Potentiodynamic polarisation was performed for the three beryllium grades in various chloride concentrations, from 0.0001 M to 0.1 M KCl solutions. Typical polarisation curves for all the beryllium grades are shown in Figure 14. This figure shows that the pitting potential of S65-Bar at -745mV SCE is higher than those for P10-Bar (-790 mV SCE) and S65-Billet (-765 mV SCE) in a 0.0001 M KCl solution. Similar differences in polarisation behaviour were observed for all the grades in 0.001M and 0.01 M KCl solution. The results from the polarisation curves for all the three grades are summarised in Table 3. This table displays the average values of open circuit potential (OCP), pitting potential (E_{pit}), repassivation potential (E_{repass}), passive current density ($i_{passive}$) and passive range (E_{pit} - OCP).

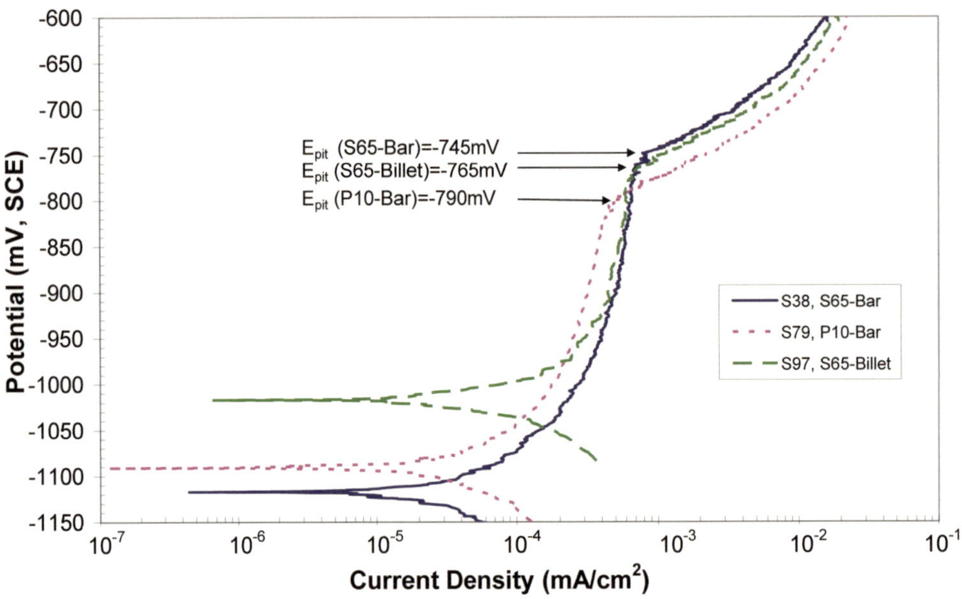

Fig. 14. Typical polarisation curves for 3 beryllium grades in 0.0001M KCl solution, sweep rate of 20mV/min

The surface condition of samples (from three beryllium grades) after polarisation in 0.001M KCl solutions was visualised. It was observed that the extent of pitting corrosion correlated well with the pitting potentials in the corresponding polarisation curves i.e. the higher the pitting potential, the lower the amount of corrosion pitting occurring. It was generally observed that polarisation at the slower sweep rate results in a denser population of corrosion pits.

Solution	Beryllium Specification	Open Circuit Potential, mV SCE	Pitting Potential, mV SCE	Repassivation Potential, mV SCE	Passive Current Density, μA/cm2	Passive Range, mV
0.0001M KCl	S65-Bar	-1068 ± 67	-750 ± 7	-802 ± 10	0.51 ± 0.08	318
	S65-Billet	-1032 ± 22	-754 ± 16	-811 ± 17	0.46 ± 0.16	278
	P10-Bar	-1033 ±81	-773 ± 25	-831 ± 1	0.34 ± 0.03	260
0.001M KCl	S65-Bar	-974 ± 66	-816 ± 21	-922 ± 26	0.31 ± 0.21	158
	S65-Billet	-934 ± 30	-835 ± 22	-926 ± 22	0.38 ± 0.22	99
	P10-Bar	-937 ± 17	-838 ± 25	-930 ± 16	0.28 ±0.20	99
0.01M KCL	S65-Bar	-1029 ± 36	-898 ± 21	-977 ± 7	0.36 ± 0.25	131
	S65-Billet	-1028 ± 17	-919 ± 2	-975 ± 9	0.40 ± 0.14	109
	P10-Bar	-1020 ± 28	-909 ±16	-962 ± 11	0.34 ± 0.11	111
0.1M KCl	S65-Bar	-1022 ± 42	-978 ± 4	-1000 ± 4	0.69 ± 0.35	44

Table 3. Pitting corrosion data from potentiodynamic polarisation curves in deaerated potassium chloride solution at pH 7.0

Figure 15 shows graphically how the pitting potentials (as listed in Table 3) for the three beryllium grades vary with chloride concentration. In all the three cases the onset of pitting potential (E_{pit}) was found to decrease logarithmically with increasing chloride concentration.

The overall relationship for pitting potential variation with chloride concentration is represented by the relationship:

$$E_{pit} = -72 \log [Cl^{-1}] - 1044 \tag{6}$$

Where E_{pit} is in mV vs SCE and $[Cl^{-1}]$ is the chloride concentration in molarity. This relationship is very close to that observed by Hill et al. (1998) for beryllium grade S200D (see Equation 4).

Moreover, the graph (Figure 15) shows that the pitting potential for S65-Bar is higher at all chloride concentrations i.e. the pitting potential decreases in the order S65-Bar > P10-Bar ≥ S65-Billet. These results suggest that S65-Bar is relatively more resistant to pitting corrosion. This work revealed that various inclusions in beryllium metal, such as intermetallic Al/Fe/Be, elemental silicon and some carbide phases act as pit initiation sites. It is also evident that there is a higher level of the coarse inclusions present in P10-Bar grade material (especially the mixed carbide and elemental silicon); and a larger sized Al/Fe/Be inclusions in the S65-Billet grade than in the S65-Bar grade. In P10 this is undoubtedly due to the significantly higher levels of C, Fe, Al, Mg and Ti impurities present (see Table 1). In the S65-Billet grade this may partly be due to the relatively higher levels of Fe and Al impurities (Table 1), and partly due to the prolonged heat treatment this material underwent during fabrication, leading to the diffusion and coalescence of impurities to form relatively large sized Al/Fe/Be inclusions at the grain boundaries. This difference in impurity content between the grades tends to explain the observed lower pitting potentials and hence the increased propensity for pitting corrosion in P10-Bar and S65-Billet grades compared to the S65-Bar grade.

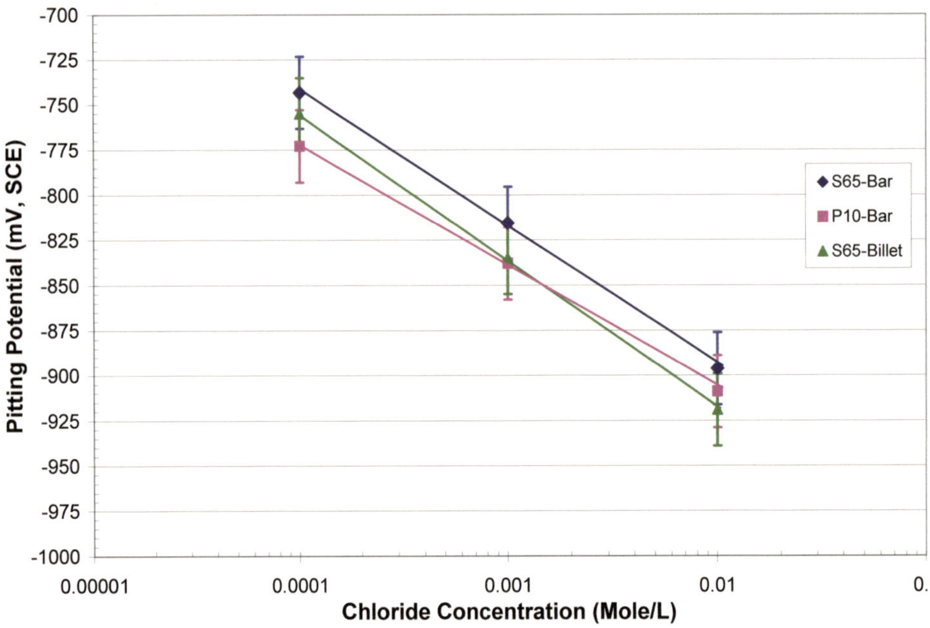

Fig. 15. Comparison of pitting potential for 3 beryllium grades at varying chloride concentration

Both pitting potential and open circuit potential can be influenced by slight variations in surface finish, solution temperature and pH value. An alternative approach was adopted by Corlett (Corlett, 2008) that the gap between the pitting potential and the open circuit potential (i.e. passive range) can be a better measure, such that the larger the passive range, the greater the resistance to pitting corrosion. Comparison of the passive ranges among all the three grades (as listed in Table 3) at various chloride concentrations also confirms that S65-Bar is more resistant to pitting corrosion than the other two grades.

There is no precise way to relate the pitting corrosion response to the extent of Al/Fe/Be inclusions in beryllium, however, as can be seen in Table 1, the level of iron and aluminium content is progressively higher in S65-Billet and P10-Bar, which can enhance the extent of Al/Fe/Be inclusions. The passive ranges, as determined in 0.0001M KCl solution, of the three beryllium grades were plotted against their aluminium content (wt%), as shown in Figure 16. This presents a tentative correlation of the pitting corrosion response to the extent of Al/Fe/Be inclusions in beryllium metal.

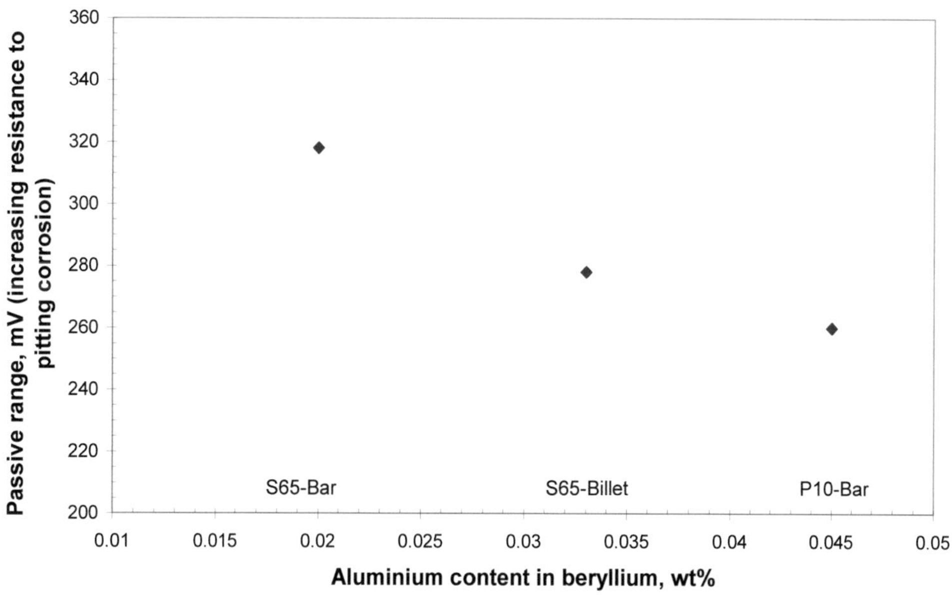

Fig. 16. Variation of pitting corrosion response in relation to aluminium content (hence the extent of Al/Fe/Be inclusion) in beryllium metal

4. Conclusions

Prior to the present experimental work being carried out, a brief review of past work on pitting corrosion of beryllium was conducted and the following conclusions were made:

1. Beryllium readily forms an adherent protective oxide film on its surface and the metal presents no corrosion problems in dry air or in high purity water at low temperature. However, it has been known that beryllium is susceptible to pitting corrosion in the presence of chloride, fluoride and sulphate ions. This is due to the breakdown of passive film at localised sites resulting in sporadic pits.
2. Electrochemical studies in NaCl solution of varying chloride ion concentrations (from 0.0001 to 1M) and pH values (2 to 12.5 pH), showed that on increasing the chloride concentration, the pitting potential decreased and at 1M NaCl, it underwent pitting corrosion at open circuit conditions. On the other hand there was an increase in pitting

potential with increase in pH. Beryllium was found to be suffering from pitting corrosion in the pH range from 2 to 12.5 but was susceptible to general attack below a pH of 2.

3. Pitting experiments on single crystals have indicated that beryllium corrodes preferentially along certain orientations. The pitting potential, E_{pit}, was found to decrease with surface orientation in the order (0001) > (1010) > (1120). In polycrystalline beryllium there was no correlation between pit initiation sites and crystallographic orientation of specific grains. Therefore, altering the bulk texture of polycrystalline beryllium will not affect the pitting corrosion resistance of beryllium, although it may affect pit propagation
4. A relationship has been observed between failure strength and chloride ion concentration in terms of the type of corrosion pits generated in beryllium. Fewer but deeper pits degrade the mechanical properties of the material to a greater extent than that by numerous but shallower pits.
5. Previously suggested mechanisms for pitting corrosion have been reviewed.

The present experimental work was aimed at determining the effect of inclusions on pitting corrosion behaviour of beryllium. The following conclusions were drawn:

6. SEM examination of the polished samples of the three beryllium grades showed that there was a presence of fine inclusion particles (0.25 to 5 μm) and some relatively coarse particles (5 to 18 μm) located along the grain boundaries. In all cases, the majority of the fine inclusions were identified as BeO and intermetallic phases such as Al/Fe/Be (and in a few cases Si/Ti or Si/Al). The coarse inclusions were elemental silicon and mixed beryllium carbides. Other elements such as Mg, Ti, U and Cr were also associated with some intermetallic and mixed carbide phases.
7. SEM examination revealed that the number of coarse inclusions (and hence the overall inclusion population) was much greater in P10-Bar than that in the other two grades. In the case of S65-billet, although the overall inclusion content was approximately the same as in the S65-bar, the intermetallic Al/Fe/Be inclusion content was much higher (i.e. larger particle size).
8. From the pit initiation study (using both electrochemical and ageing techniques), it was observed that corrosion pits were nearly always initiated at the metal grain boundaries. Although the intermetallic Al/Fe/Be and elemental silicon inclusions were the preferential sites for pit initiation, some corrosion pits had also initiated at carbide inclusions.
9. Electrochemical polarisation of the three beryllium grades showed that the pitting potential of S65-Bar grade was relatively high compared to the other two grades. This is in agreement with the extent of corrosion pitting revealed on the coupons after polarisation.
10. The nature and extent of inclusions in each grade showed a strong relationship with its pitting potential. The lower overall inclusion content of the S65-Bar compared to the P10-Bar, and its lower Al/Fe/Be intermetallic inclusion content compared to the S65-Billet, resulted in it displaying a corresponding higher pitting potential and hence a relatively greater resistance to pitting corrosion.
11. It is evident from the present work that the various inclusions in beryllium located at the grain boundaries, such as Al/Fe/Be, elemental silicon and carbides act as pit initiation sites and hence their presence is likely to enhance the propensity for pitting corrosion.

5. Acknowledgements

The author would like to thank Ms. Jennifer Copeland of Brush Wellman Industries for the chemical analysis. The author also wish to acknowledge Mr Mike Cox, Dr Andrew Wallwork and Dr S McCulloch of AWE for their support.

6. References

Blanc, C. & Mankowski, G. (1997). Susceptibility to Pitting Corrosion of 6056 Aluminium alloy, Corrosion Science, vol. 39, 1997, pp. 949-959

Carrabine, J. A. (1963). Ternary $AlMBe_4$ Phases in Commercially Pure Beryllium, J. Nuclear Materials, Vol. 8, pp. 278-280.

Corlett, N. (2008). Corrosion Checking, Materials World 2008, Vol. 16, pp. 27-29

Friedman, J. R. & Hanafee, J. E. (2000). Corrosion/Electrochemistry of Monocrystalline and Polycrystalline Beryllium in Aqueous Chloride Environment, UCRL-ID-137482

Flitton, M. K. A.; Mizia, R. E. & Bishop, C. W. (2002) Understanding Corrosion of Activated Metals in an Arid Vadose Zone Environment, INEEL/CON-01-01450, NACE

Gulbrandsen, E. & Johansen, A. M. J. (1994), A Study of the Passive Behaviour of Beryllium in Aqueous Solutions, Corrosion Science, Vol. 36, pp. 1523-1536

Hill, M. A.; Butt, D. P. & Lillard, R. S. (1996). The Corrosion/Electrochemistry of Beryllium and Beryllium Weldments in Aqueous Chloride Environments, Los Alamos National Laboratory New Mexico, Internet Report No. 87545.

Hill, M. A.; Butt, D. P. & Lillard, R. S. (1998). The Passivity and Breakdown of Beryllium in Aqueous Solutions, J. Electrochemical Soc., Vol. 145, pp. 2799-2806

Kaye, G. W. C. & Laby T. H. (1978). Tables of Physicals and Chemical Constants, Longman Press, New York 1978, p. 216

Lillard, R. S. (2000). Factors Influencing the Transition from Metastable to Stable Pitting in Single-Crystal Beryllium, J. Electrochemical Soc., Vol. 148, pp. B1-B11

Miller, P.D. & Boyd, W. K. (1968). Beryllium Deters Corrosion- some do's and don'ts, Materials Engineering (July 1968) pp. 33-36

Mueller, J. J. & Adolphson, D. R. (1979). Corrosion, Beryllium Science and Technology 2, D R Floyd and J N Lowe (eds), Plenum Press, New York 1979, pp. 417-432

Mathiesen, T. &. Hansen, J. V. (2010). Consequences of Sigma Phase on Pitting Corrosion Resistance of Duplex Stainless Steel, Duplex World Oct 2010 Conference, Beaune, France

Pourbaix, M. (1966). Atlas of Electrochemical Equilibria in Aqueous Solutions, Pergamon Press, New York, p. 135

Pride, S. T.; Scully, J. R. & Hudson, J. L. (1994). Metastable Pitting of Aluminium and Criteria for the Transition to Stable Pit Growth, J. Electrochemical Soc., Vol. 141, pp. 3028-3040

Punni, J. S. & Cox, M. J. (2010). The Effect of Impurity Inclusions on the Pitting Corrosion Behaviour of Beryllium, Corrosion Science, Vol. 52, pp. 2535-2546

Rooksby, H. P. (1962). Intermetallic Phases in Commercial Beryllium, J. Nuclear Materials, Vol. 7, pp. 205-211

Stonehouse, A. J. & Weaver, W. W. (1965). Beryllium Corrosion, Materials Prot., Vol. 4, pp. 24-36

Szklarska-Smialowska, Z. (1999). Pitting Corrosion of Aluminium, Corrosion Science, Vol. 41, pp. 1743-1767

Vaidya, R. U.; Hill, M. A.; Hawley, M. & Butt, D. P. (1998). Effect of Pitting Corrosion in NaCl Solutions on the Statistics of Fracture of Beryllium, Metallurgical and Materials Transactions A, Vol. 29A, pp. 2753-2760

Vaidya, R. U.; Brozik, S.M.; Deshpande, A.; Hersman, L. E. & Butt, D. P. (1999). Protection of Beryllium Metal against Microbial Influenced Corrosion using Silane Self-Assembled Monolayers, Metallurgical and Materials Transactions A, Vol. 30A, pp. 2129- 2134

Venugopal, A.; Macdonald, D. D. & Verma, R. (2000). Electrochemistry and Corrosion of Beryllium in Buffered and Unbuffered Chloride Solutions, J. Electrochem. Soc., Vol. 147, pp.3673-3679

West, J. M. (1970). Electrodeposition and Corrosion Processes, Plenum Press, New York, p. 94

2

Mechanism of Pit Growth in Homogeneous Aluminum Alloys

G. Knörnschild
Federal University of Rio Grande do Sul
Brazil

1. Introduction

Pitting corrosion is a process, which takes place on passive metals and alloys. A characteristic of this type of corrosion is that passivity breaks down at isolated points at the surface and the growth of pits is observed due to locally high rates of metal dissolution. In electrochemical experiments, the growth of pits leads to a rapid rise of the overall current density once a characteristic threshold potential, the pitting potential E(pit) is surpassed. Since the measured current density is composed of the passive current density at the passive surface area and the current density of fast metal dissolution at the pitted area conventional electrochemical tests are not useful for studying metal dissolution inside pits. Some authors tried to overcome this difficulty by working with small wire electrodes. The idea behind was to achieve an electrode state where the whole surface represents a pit and the measured current density becomes, therefore, identical to the real current density inside a pit. However, highly concentrated electrolytes and high potentials must usually be applied to achieve this electrode state [1-3]. Formation of salt films and mass transport control was usually observed under these conditions, which more likely represent the conditions of electropolishing rather than that of a metal suffering pitting corrosion at the corrosion potential.

Other authors determined the current density during the initial stage of pit growth from the microscopic measurement of the dimensions of pits formed in short time experiments [4]. By measuring the time for perforation by pitting of thin metal foils Hunkeler [5] and Cheung [6] obtained average rates of pit growth, i.e., average current densities normal to the foil surface. Average rates of localized corrosion at grain boundaries of aged AlCu alloys have been determined by metallographic measurements of penetration depth and penetration time [7]. By the same method early stages of pit propagation were studied in 7075-T3 alloy [8]. Few studies have been made to measure in-situ metal dissolution inside pits. Edeleanu [9] examined the pit propagation in thin aluminum foils, glued to a glass foil which could be observed from the back side by a microscope. In this way pit propagation along the glass foil could be observed in situ. The technique was applied again for an intensive study of pit growth in pure aluminum by Baumgärtner [10-12], and for the study of homogeneous aluminum alloys by Knörnschild and Kaesche [13,14]. Later, Frankel [15] used a similar technique to study pit growth in thin aluminum films deposited by PVD.

2. Experimental techniques

Pure aluminum as well as homogeneous Al-4wt.%Cu and Al-3wt.%Zn alloys were studied in NaCl and $AlCl_3$-solutions of varying concentrations, using a conventional three-electrode-arrangement. Tests were performed either in potentiostatic, potentiodynamic or galvanostatic mode. Samples were homogenized by annealing for one hour at 480°C (Al), 530°C (AlCu) and 400°C (AlZn), respectively. After annealing the samples were quenched in cold water. The cold water quench of AlCu was interrupted for five seconds in boiling water in order to prevent accelerated clustering of copper due to quenching stresses.

The experimental arrangement for the microscopic observation of pit growth is shown in Fig.1. The thin samples were glued to a glass foil which formed the bottom of the electrochemical cell. The lower side of the sample could be observed by a microscope and filmed with a coupled camera. The samples were polarized potentiostatically or galvanostatically in chloride solution. In samples with (100)-orientation metal dissolution during pitting occurs perpendicular to the glass foil surface. Once a pit has grown enough to reach the glass foil the current density of metal dissolution can be determined by image analysis from the shift of the tunnel front.

Tunnels in AlZn alloy were further studied by an oxide replica technique; i.e., the pitted samples were anodized during 60s at 70V in $NH_4H_2PO_4$-electrolyte and the metal was chemically dissolved in a methanolic Br_2 solution, leaving only the anodic oxide film.

The current densities of hydrogen evolution during pitting were determined volumetrically by the collection of the evolved H_2 gas.

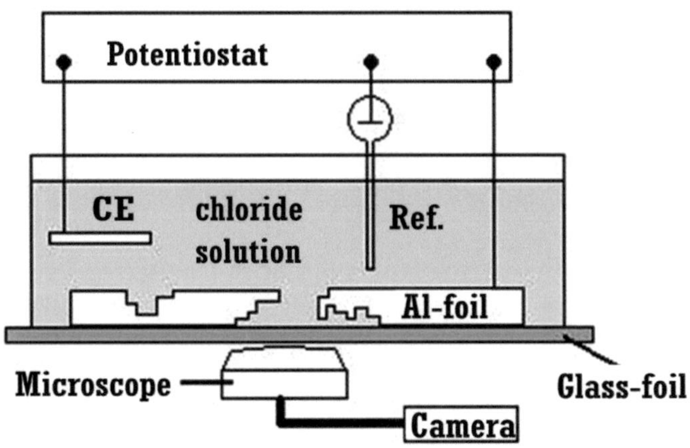

Fig. 1. Experimental arrangement for microscopic observation of pit growth.

3. Results

3.1 Morphology of pitting attack

3.1.1 Al and AlCu-alloy

Pure aluminum and the homogeneous aluminum alloys have a pronounced pitting potential E(pit) in 1M NaCl solution. Stationary potentiostatic or galvanostatic measurements show that the electrodes become nearly unpolarizable at this threshold potential (Fig.2). Metal dissolution during pitting in aluminum and in homogeneous aluminum alloys at or above this threshold potential leads to the formation of crystallographic pits. In the AlCu alloy, pits grown in chloride solution have walls, which consist of (100)-planes (Fig.3). In principle, the morphology is identical with that observed in pure aluminum. However, while pure aluminum pits shows sharp (100)-planes (Fig.4), rounded walls are more frequently found in the AlCu alloy (Fig.3). Transmission electron microscopy of pitted samples revealed that these rounded walls consist of fine crystallographic (100)-steps (Fig.5). It was also observed that more corrosion products, frequently containing Cu, are found at the surface of pits in AlCu than in pure Al.

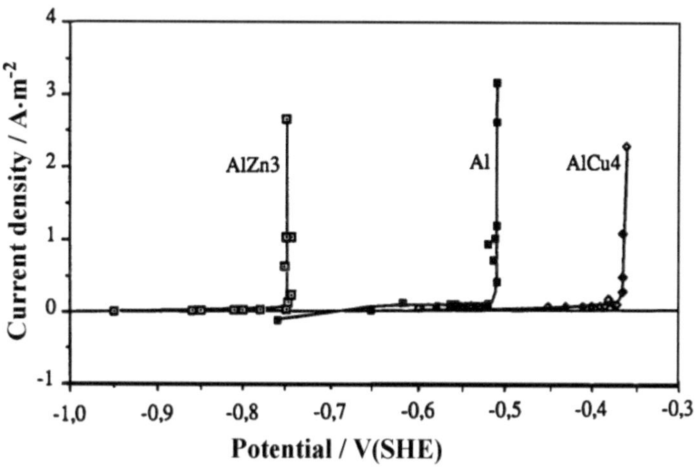

Fig. 2. Stationary current density – potential curve of Al, homogeneous Al-4wt.%Cu and homogeneous Al-3wt.%Zn in deaerated 1M NaCl solution.

Fig. 3a. Morphology of pits in the the AlCu alloy in 1M NaCl.

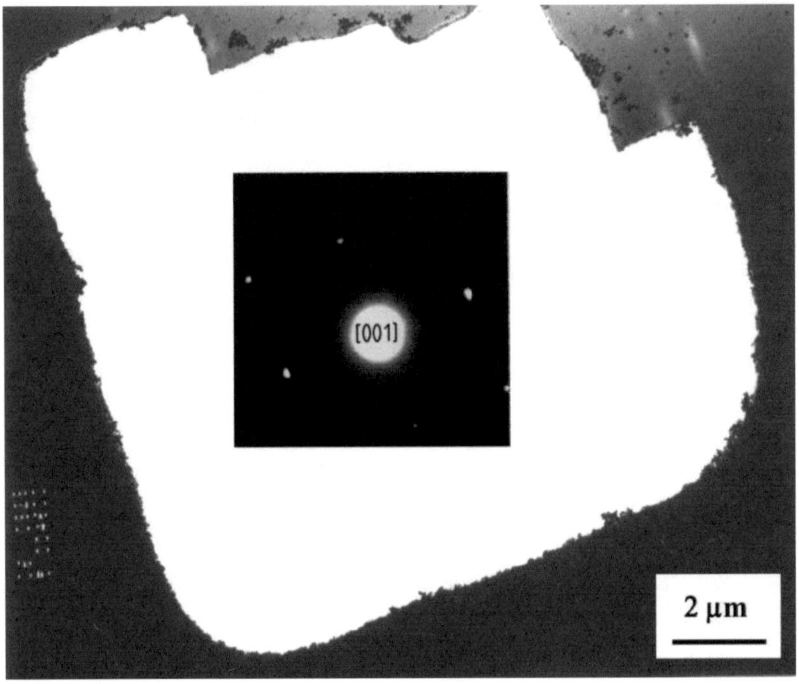

Fig. 3b. Transmission electron micrograph of pit in the AlCu-alloy with corresponding electron diffraction pattern showing (100) pit walls.

Fig. 4. Morphology of pits in pure Al in 1M NaCl.

Fig. 5. Transmission electron micrograph of AlCu-alloy showing crystallographic fine structure of rounded pit walls.

3.1.2 AlZn-alloy

In the homogeneous AlZn alloy tunnels with rounded cross section grow in crystallographic directions (Fig.6). Growth directions have been confirmed as <100> directions by electron diffraction of pitted TEM samples. Tunnel diameters are up to 0,5µm. The morphology is very similar to tunnels which grow in pure aluminum in hot hydrochloric acid. These conditions are widely used for the etching of capacitor foils [16]. The tunnels in capacitor foils, however, have a quadratic cross section, i.e., the tunnel side walls are crystallographic.

Fig. 6a,b. Pitting attack in AlZn alloy in 1M NaCl, showing accumulation of tunnels.

In the case of tunnels in AlZn only the growth direction is crystallographic, while the side walls do not show any crystallographic aspects, as confirmed by TEM examinations (Fig.7). When tunnels in AlZn are produced in acid chloride solution they show crystallographic cross sections like capacitor foils, even when produced at room temperature (Fig.8). A drawback for the industrial use of the AlZn tunnels is that they show a strong tendency to clustering (Fig.6 and 9) instead of a uniform tunnel distribution. A great number of parallel tunnels usually starts to grow from the same point of attack (Fig.9).

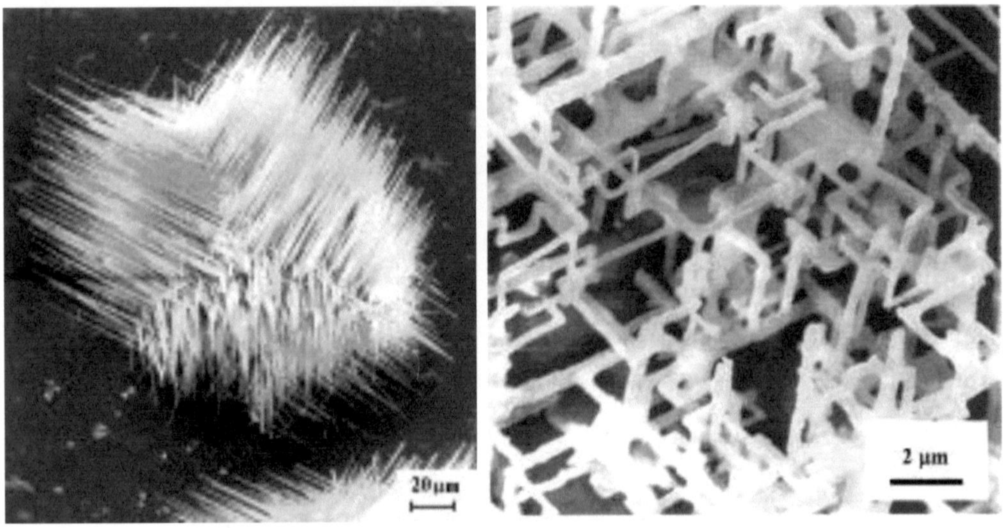

Fig. 6c,d. Tunnels in AlZn alloy in 1M NaCl, revealed by oxide replica technique.

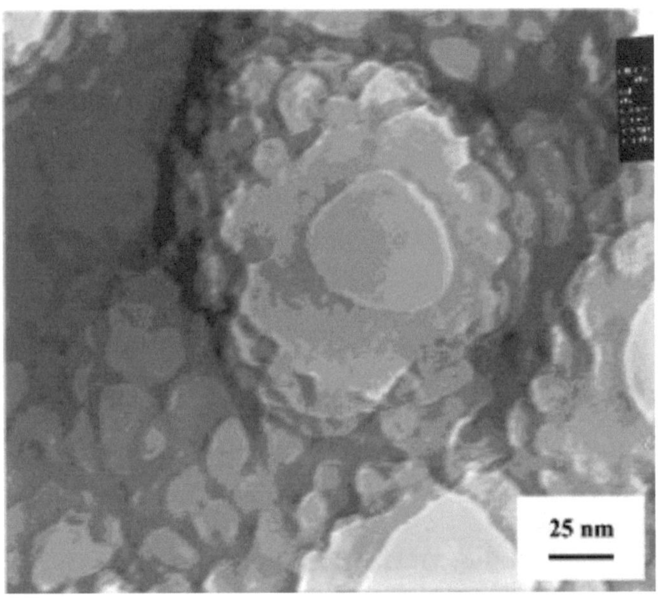

Fig. 7. Transmission electron micrograph of tunnel walls in AlZn-alloy.

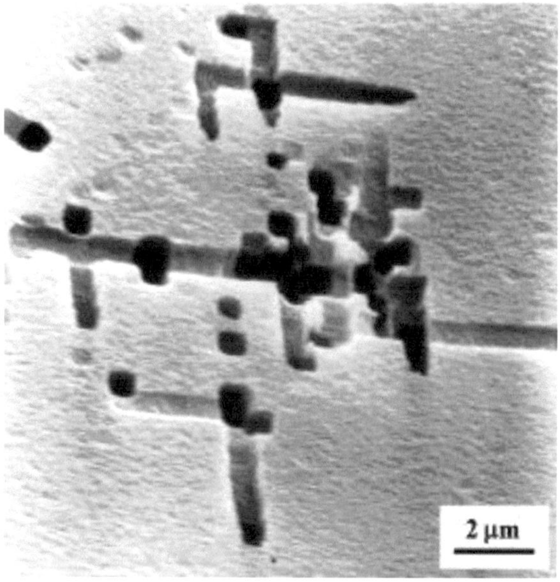

Fig. 8. AlZn: Tunnels with quadratic cross section grown in solution of NaCl, pH1,5.

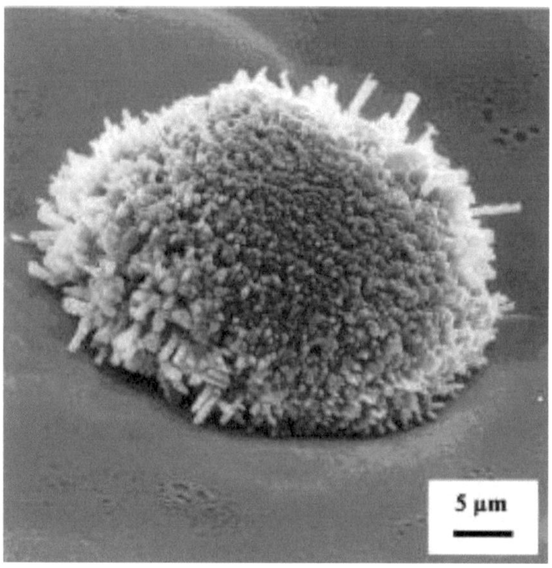

Fig. 9. AlZn-alloy: Early stages of tunnel growth in 1M NaCl solution.

3.2 Pit propagation studied by in-situ microscopic observation

3.2.1 AlCu-alloy

The growth of pits occurs by subsequent dissolution of crystallographic steps at (100)-planes. By periodic restriction of the area of the dissolving planes a tunnel-like morphology is formed, in the example shown in Fig.10 this tunnel propagates approximately in <110>-direction.

The life time of localized dissolution events, such as that shown in Fig.10 is limited, even at or above the pitting potential. Repassivation may occur after some seconds or localized dissolution may last up to about one minute. At higher applied potentials life time tends to be longer. While one tunnel repassivates, other tunnels begin to propagate. This means that macroscopic pits are formed by a great number of localized, short-lived dissolution events, occurring simultaneously and subsequently. At any time, only a small part of the inner surface of a macroscopic pit is actively dissolving. The formation of macroscopic pits probably happens due to the fact that in front of recently repassivated surfaces the electrolyte is enriched in aluminum and chloride ions. This environment with low pH and high concentration of passivity destroying ions makes the nucleation of new tunnels more probable in comparison to nucleation at the passive surface outside the pit.

From the shift of the crystallographic planes the average current density during pit growth was calculated by Faraday's law. The current densities, obtained from testes in diluted NaCl solutions vary between $3A/cm^2$ and $10A/cm^2$ (Fig.11). In the case of pure aluminum, the same average current densities were obtained in diluted NaCl solution [10-12], however, up to $30A/cm^2$ were measured during short time intervals, proving that the dissolution of the crystallographic fronts indeed is a discontinuous process. In the case of the homogeneous AlCu alloy signs of discontinuous propagations were also found. However, discontinuities seem to be at a finer scale and are therefore close to the spatial resolution of the microscope.

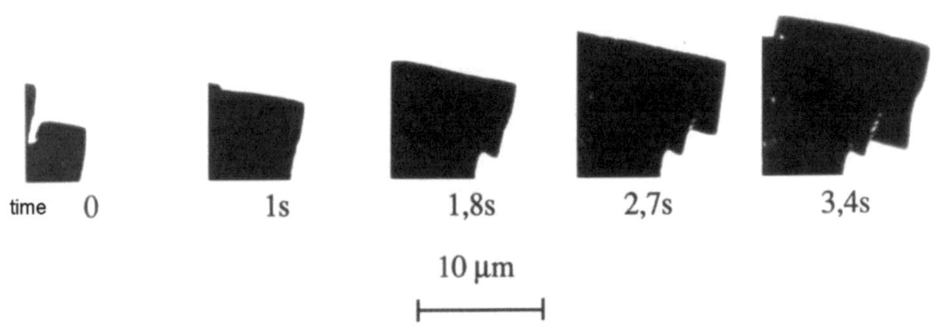

Fig. 10. Microscopic in-situ observation of tunnel growth in AlCu-alloy.
Electrolyte: 1M NaCl.

Average current densities remain almost constant during tunnel propagation, as can be seen in Fig.11. In no case the current density depended on the pit depth or tunnel length. It was also confirmed that chloride concentration (except saturated $AlC_3{\cdot}6H_2O$ solution, as described below), as well as applied potential above E(pit) and applied current density in galvanostatic tests have no influence on the measured average current density of actively dissolving tunnel fronts.

Dissolving tunnel fronts stopped immediately, i.e., within tenths of a second, when the potential was switched below the pitting potential E(pit) indicated in the stationary current density-potential-curve. The behavior of growing tunnels during potential switch experiments to E < E(pit) has been intensively studied for pure aluminum in NaCl solution by Baumgärtner [10-12]. It was found that the time to repassivation was shorter the lower the potential E < E(pit) to which the potential was switched. The tests fulfilled with the homogeneous AlCu alloy indicated that the behavior of the alloy is qualitatively the same.

The only influence on the current density of growing tunnel fronts was found in highly concentrated or saturated $AlCl_3{\cdot}6H_2O$ solution. In this case the shift of broad dissolution fronts instead of rather large (100) planes is observed. Ex-situ electron microscopy revealed that the dissolution front in concentrated or saturated $AlCl_3{\cdot}6H_2O$ solution has also a fine structure of (100) steps. The average current density of aluminum dissolution under these conditions was in the order of 0,3A/cm², which means one order of magnitude below that in diluted chloride solutions. Here, again, pure aluminum and homogeneous AlCu alloy showed very similar results.

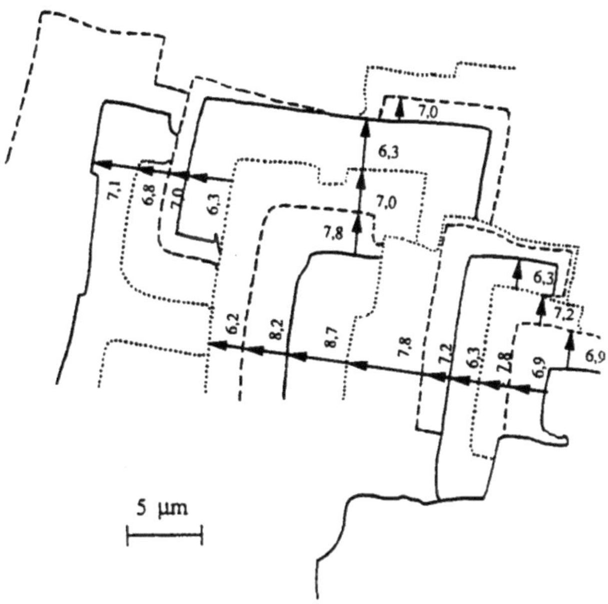

Fig. 11. Propagation of tunnel front in AlCu alloy, obtained from in-situ observations. Electrolyte: 1M NaCl. Time interval between two lines: one second; numbers indicate current density in A/cm².

3.2.2 AlZn-alloy

In the homogeneous AlZn alloy one-dimensional growth of tunnels with cylindrical cross section of some tenths of µm in crystallographic directions is observed (Fig.6). The tunnel diameter remains constant over the tunnel length, which means that dissolution of the tunnel walls is negligible in comparison to the dissolution at the tunnel tip.

As in the case of pure aluminum and the AlCu alloy the rate of metal dissolution, in this case the propagation of the tunnel tip, did not depend on tunnel length, as can be seen in Fig.12. Current densities at the tunnel tip, obtained from this and from other experiments in diluted NaCl solution were about 5A/cm². The in-situ microscopic technique is not very appropriate for the observation of the one dimensional tunnels in the AlZn alloy, since it needs a great number of experiments to find tunnels which propagate exactly in the plane of the glass foil. Therefore additional short time experiments were made with conventional samples. The average current density of tunnel propagation was calculated from the tunnel length, measured by the oxide replica technique and the time of tunnel growth (being supposedly equal to the duration of the experiment). Average current densities between 2 and 7A/cm^2 were obtained in this way. This means, that the values are in the same order of magnitude as in the case of aluminum and of the AlCu alloy. The tunnel life in AlZn alloys is also limited. The maximum tunnel length registered in the all the tests performed with the AlZn alloy was about 100 μm (Fig.6a,c).

Again, as in the case of aluminum and of the AlCu alloy, chloride concentration, applied potential or applied current density in galvanostatic tests have no influence on the current density. The only condition, where different results are obtained, is again saturated $AlCl_3 \cdot 6H_2O$ solution: In the case of the AlZn alloy the tunnel length in saturated $AlCl_3 \cdot 6H_2O$ solution is shorter than in diluted chloride solutions, i.e. tunnel life times are shorter. Because of the very short life time of these tunnels the current densities could not be determined under these conditions.

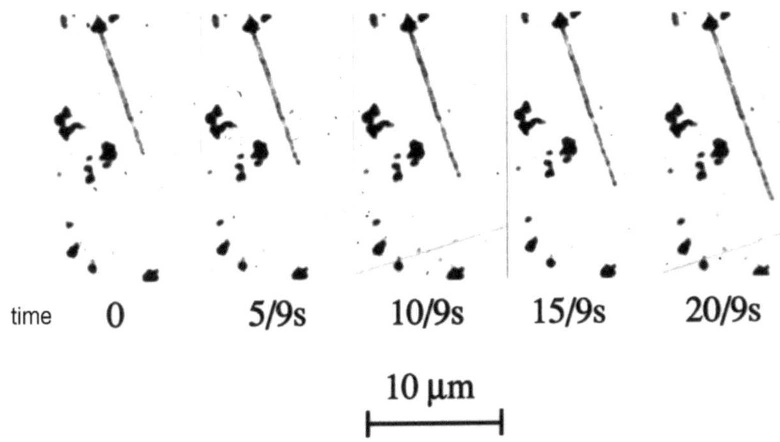

Fig. 12. Microscopic in-situ observation of tunnel growth in AlZn. Electrolyte: 1M NaCl.

3.3 Hydrogen evolution as a partial reaction during pitting

Nucleation of H_2 bubbles can be observed during pitting in pure aluminum and in the AlCu alloy. By microscopic in situ observations it becomes clear that hydrogen evolution is spatially correlated with sites of fast active metal dissolution inside pits, however, this doesn't mean that bubble formation occurs directly at the active pit surface. In the fine tunnels in AlZn bubble nucleation cannot take place during tunnel growth, since bubble formation would block completely the potential control of the tunnel front. This is also confirmed by microscopic observations. During tunnel growth formation of bubbles was not observed inside the tunnels. However, sometimes bubbles formed inside a tunnel after its repassivation. From these observations it becomes clear that mass transport inside growing AlZn tunnels is not accelerated by hydrogen bubbles and that blockage of potential control at the tunnel tip by bubbles is not the reason for repassivation.

Volumetric measurements of hydrogen evolution acompanying pitting corrosion showed a chacteristic value of i(H)/i(Al), without considerable dependance of the chloride concentration (Fig.13). No appreciable change was observed when tests were performed in diluted iodide solutions. Also, the differences between Al, AlCu and AlZn were small. In all the performed tests the ratio i(H)/i(Al) remained in the interval between 0,09 an 0,17. The variation of the pitting potential E(pit) in these tests was almost 800mV (Fig.13). Consequently, i(H)/i(Al) does not depend on the electrode potential.

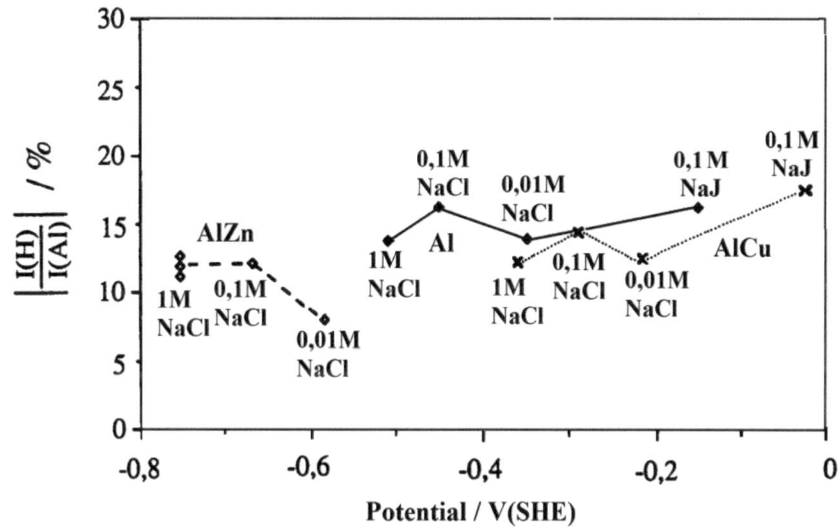

Fig. 13. Relation of current densities of hydrogen reduction i(H) to Al-dissolution during pitting i(Al). Volumetric measurements with Al, AlCu and AlZn in chloride and iodide solutions.

4. Discussion and conclusions

The present work has shown that macroscopic pitting is the result of the growth of isolated tunnels. Each one of these tunnels has a limited lifetime. At any time only a small area is active within a macroscopic pit. This behavior explains why current densities measured in-situ at active tunnel fronts are much higher than current densities obtained from the penetration time by pitting of thick specimens. Even in penetration experiments with thin foils as those performed by [5,6] penetration times may be longer than single short-lived active dissolution events which can be observed by microscopic in-situ experiments. In this case, current densities, calculated from penetration times, are averaged values which contain periods of rapid local dissolution and times of standstill. Therefore, one has to be careful with conclusions about dissolution kinetics, which were drawn, for example, from the potential dependence or and the thickness dependence of the results.

In situ pit growth experiments have shown that the current density of Al dissolution at actively dissolving pit fronts is practically independent of the alloy composition for pure Al, homogeneous AlZn and AlCu alloy, as well as independent of the externally applied electrode potential. This has been confirmed in tests with the different alloys and with different chloride concentrations, in which the potential variation was more than 0.5V. Current density of metal dissolution, when measured in time intervals of some tenths of a second always remained in the range of 3 - 10A/cm^2. The independence of metal dissolution rate from applied electrode potential could be interpreted as mass transport control. However, from the fact that the dissolution rate of active pit fronts depends neither on the tunnel length nor on the concentration of Al^{3+} and Cl^- in the bulk electrolyte transport control of Al dissolution during pitting can be ruled out. Volumetric tests showed that the rate of hydrogen evolution accompanying pitting of Al, AlCu and AlZn is related to the rate of Al dissolution. I.e., the relation of i_H/i_{Al} varies between 0.09 and 0.17, independent of externally applied current or potential, chloride or iodide concentration. This was found for the current densities related to the geometrical surface of the samples, but certainly this also holds for the true local current densities I_H and I_{Al} inside the pits. The volumetric tests were performed within a potential range of about 0.8V, so the relation of I_H/I_{Al} can be considered independent of the applied potential.

The results about Al dissolution and hydrogen evolution show that the two reactions are coupled. They do not represent independent partial reactions. From these results it becomes obvious that at the active pit front always the same conditions are maintained, characterized by a typical current density of Al dissolution of 3-10A/cm^2, and a current density of hydrogen evolution which is between 9 and 17% of the anodic current density. This holds independent of the set of external parameters: applied current density, applied potential, kind and concentration of halide ion and Al^{3+} in the bulk electrolyte; nearly saturated $AlCl_3{\cdot}6H_2O$ solution being the only exception.

The fact that Al dissolution in highly concentrated $AlCl_3{\cdot}6H_2O$ solutions with a pH $\approx$0 is considerably slower than dissolution in neutral bulk NaCl solution shows that acidification alone is not the critical factor which allows the special kind of high rate metal dissolution occurring during pitting of Al. It rather points to the existence of a metastable surface film which permits metal dissolution into an electrolyte which is normally not saturated with respect to the film forming ions. In principle this is the same situation which is found in most cases of passivity. The passive film is usually thermodynamically unstable, since the electrolyte is not saturated with respect to the passive film. In practice, most passive films are protective because of the slow dissolution kinetics not because of its thermodynamic stability. In the case of pitting the only difference is the high rate of film formation and dissolution.

Although further information about the exact nature of the film is not available, it is clear that the active Al surface must be covered by chloride or some oxychloride species, since otherwise Al would passivate in the aqueous environment. The film permits the fast passage of Al ions and can consequently be considered an $AlCl_3$-film. Kaesche [14] describes this salt film as $AlCl_z{\cdot}xH_2O$, which means that its exact stoichiometry and its degree of hydration are not known. Beck [1] concluded from experiments with wire electrodes that an unhydrous film exists at the metal surface and that this film is hydrated from the electrolyte side. Beck found that at sufficiently high potentials and in highly concentrated $AlCl_3$ electrolyte salt film dissolution becomes mass transport controlled. Under these conditions the film thickens and water diffusion into the film to the metal-film interface determines the rate of hydrogen reduction. With growing potential and thus growing film thickness the ratio I_H/I_{Al} diminishes. The same observation Beck made for mass transport controlled dissolution of Mg [3].

This model means that hydrogen is reduced at the metal-film interface and that reduced hydrogen must diffuse back through the film to the electrolyte, where bubble formation occurs. It should be mentioned that an alternative, well known model for hydrogen evolution during pitting exists: Dissolution of Al as Al^{2+} or Al^{+} and further oxidation of these ions by H^{+} in a homogeneous reaction. The fact that in the present work a characteristic current density of Al dissolution and a characteristic ratio of I_H/I_{Al} were found in all cases might show that these conditions are indispensable for the kind of metal dissolution during pitting of Al.

Parting from Beck´s model that water is transported through the salt film, a constant ratio of I(H)/I(Al) might indicate, that the high dissolution current is necessary to hold the O^{2-}/Cl^{-} ratio (O^{2-}-formation by $H_2O + 2e^- \rightarrow 2H + O^{2-}$) at the metal surface below a critical value. Slower metal dissolution makes the O^{2-} coverage of the metal surface rise and causes repassivation. Obviously, this condition is affected when dissolution of Al becomes controlled by diffusion. Simple estimations for the Al^{3+} transport show that it is difficult to maintain current densities of some A/cm^2, when the diffusion layer reaches the order of 100µm. The assumption of repassivation by beginning transport control is sustained by various experimental observations: Tunnel length in the AlZn alloy is limited to about 100µm. Studies of tunnel growth in aluminum at about 80°C in HCl, from Alkire [16], also showed tunnels, whose length do not surpass 100µm. In another work, it was shown that the tunnel length in aluminum at 70°C diminishes with growing $AlCl_3$ concentration in the electrolyte [17]. In the present work tunnels in the AlZn alloy reached only a length of a few µm in saturated $AlCl_3{\cdot}6H_2O$ solution.

Al and AlCu in saturated $AlCl_3{\cdot}6H_2O$ solution can maintain the necessary high current density for tunnel growth only during very short time intervals, due to the slow transport between the surface with some supersaturated $AlCl_3$ film and the $AlCl_3{\cdot}6H_2O$ saturated electrolyte. Pit growth stops due to beginning transport control but restarts moments later when the concentration has declined. As a result of the discontinuous propagation the average current density of Al dissolution is lower.

Hebert and Alkire [18] also considered diffusion as the decisive factor for repassivation. According to these authors the growth of one-dimensional tunnels stops when the diffusion layer extends over the whole tunnel length and the tunnel tip becomes saturated with $AlCl_3{\cdot}6H_2O$.

Kaesche [14] suggested that the sum of diffusion potential and ohmic potential drop rises during tunnel growth and repassivation occurs when the potential at the tunnel bottom falls below some critical value. From the thermodynamic point of view the critical value is the equilibrium potential of the film forming reaction.

Potential switch experiments showed that the external potential where repassivation of growing pits occur is identical with the potential E(pit) obtained from the steep rise of the stationary current density - potential curve.

Separate pitting and repassivation potentials are found especially in works about metastable alloys [19]. These are interesting because alloys supersaturated with alloying elements like Mo, W, Ta [20-22] have much higher pitting potentials than those that can be achieved by conventional alloys like Al-Cu. Due to the fabrication process, PVD techniques or rapid solidifying these alloys are only available as very thin films or melt spun ribbons. Resent progress in producing supersaturated bulk materials is so far restricted to alloying systems based on Zr, Fe, Ni and Cu [23]. The problem with the reported pitting potentials is that, due to the thin format of the samples, measurements are generally made non-stationary by potentiodynamic tests.

The question of repassivation was sometimes also discussed because the reverse scan of potentiodynamic tests reveals that pitted samples show considerably higher current densities below E(pit) than the unpitted samples during the forward scan. Although the phenomenon is easy to observe, few authors have treated this aspect of Al corrosion [24-26]. Sometimes, the potential where the steep fall of current during the reverse scan is substituted by a less potential dependent current region is called pit transition potential [27,28]. The elevated current density can be attributed to an acidified surface film which remains in the pits formed at E>E(pit). From the potential switch experiments in the present work, which showed that growing pits stop within tenths of seconds, one has to conclude that these currents are not caused by some kind of pitting corrosion which continues below E(pit), although Moore [26] tried to apply a model from Newman [29] who postulated that pits could reduce its active surface, transforming itself into tunnels below E(pit).

Morphological studies from Moore [26] showed the formation of some tunnels as postulated by the model. However, it is widely accepted that the formation of metastable pits or tunnels below E(pit) occurs also at passive surfaces without previous pitting attack [25] and at the pitted surface with an acidified surface film the number of metastable pits or tunnels might be enhanced.

It is important to know that residual currents continue for several hours or even days when the reverse scan is stopped and the samples are held potentiostatically at E<E(pit) [30]. The time the current needs to cease depends on the charge which had passed at E>E(pit). This shows that a quasi-stationary state exists in which the acidified electrolyte in the repassivated pits and tunnels below E(pit) is maintained by diffusion and migration between the pitted surface and the neutral bulk electrolyte, thus maintaining acid attack of the surface for a long time. After the decay of the current what is found on the surface is a rounded shallow attack, instead of crystallographic pitting. The decay of current occurs because the acid attack leads to a smoothening of the surface. As a consequence the surface pH maintained by metal dissolution and transport begins to rise and the attack comes to a standstill. The attack can be considered as a kind of crevice corrosion, where the corrosion is consuming the crevice.

The decisive difference between pitting corrosion and the corrosion by residual currents below E(pit) is the state of the surface. Pitting means that the surface is maintained active by some chloride containing film. Dissolution is localized and rapidly dissolving surfaces are coexisting side by side with passive surfaces. Constant attack of this kind leads to more and more roughened surfaces. Corrosion by residual currents, on the other hand, means metal dissolution through a passive film thinned by an acid electrolyte. Dissolution occurs on the whole surface, since it is determined by local pH. This leads finally to a smoothening of the surface previously roughened by pitting. The rounded, shallow attack has the morphology which is typical for Al dissolution in an acid electrolyte.

5. References

[1] Beck T.R., Salt Film Formation during Corrosion of Aluminum, Electrochimica Acta, 29 (1984) 485-491.

[2] Beck T.R., Mueller J.H., Conductivity and Water Permeability of Barrier-Layer $AlCl_3$ as a Function of Temperature, Electrochimica Acta, 33 (1988) 1327-1333.

[3] Beck T.R., Chan S.G., Corrosion of Magnesium at High Anodic Potentials J, Electrochem.Soc. 130 (1983) 1289-1295.

[4] Streblow H.H., Ives M.B., On the electrochemical conditions within small pits, Corr.Sci. 16 (1976) 317-318.

[5] Hunkeler F., Böhni H., Determination of Pit Growth-Rates on Aluminum Using a Metal Foil Technique, Corrosion, 37 (1981) 645-650.

[6] Cheung W.K., Francis P.E., Turnbull A., Test Method for Measurement of Pit Propagation Rates, Materials Science Forum 192-194 Part1 (1995) 185-196.

[7] Wenzel G; Knörnschild G., Kaesche H, Intergranular Corrosion And Stress-Corrosion Cracking of an Aged Alcu Alloy in 1-N Nacl Solution Werkstoffe und Korrosion-Materials and Corrosion 42 (1991) 449-454.

[8] .Glenn A.M, Muster T.H., Luo C., Zhou X., Thompson G.E., Boag A., Hughes A.E., Corrosion of AA2024-T3 Part III: Propagation, Corrosion Science 53 (2011) 40-50.

[9] Edeleanu C., The Propagation of Corrosion Pits in Metals, J. Inst. Metals 89 (1960) 90-94.

[10] Baumgärtner M., Kaesche H., Microtunnelling During Localized Attack of Passive Aluminum - the Case of Salt Films vs Oxide-Films Corr. Sci. 29 (1989) 363-378.

[11] Baumgärtner M., Kaesche H., Aluminum Pitting in Chloride Solutions - Morphology and Pit Growth-Kinetics Corr. Sci. 31 (1990) 231-236.

[12] Baumgärtner M., Kaesche H., The Pitting Potential of Aluminum in Halide Solutions; Werkstoffe und Korrosion 42 (1991) 158-168.

[13] Knörnschild G., Kaesche H., Localized corrosion of homogeneous binary Al-Zn and Al-Cu alloys. The Electrochemical Society Fall Meeting, Toronto, Oct.11-16, 1992, Vol.92-2, p. 178.

[14] Kaesche H.: Corrosion of Metals, Springer, 2003, Chapter 12.

[15] Frankel G.S., Pit growth in thin metallic films, Mat. Sci. For., 247 (1997) 1-7.

[16] Alwitt R.S., Uchi H., Beck T.R., Alkire R.C., Electrochemical Tunnel Etching of Aluminum, J.Electrochem.Soc 131 (1984) 13-17.

[17] Alwitt R.S., Beck T.R., Hebert K., Proc. Int. Conf. Localized Corrosion, Orlando 1987, Ed. H.S.Isaacs, NACE, Houston, p.145.

[18] Hebert K.R., Alkire R., Growth and Passivation of Aluminum Etch Tunnels, J.Electrochem.Soc.135 (1988)2146-2157.

[19] Lucente A.M., Scully J.R., Pitting and Alkaline Ddissolution of an Amorphous-Nanocrystalline Alloy with Solute-Lean Nanocrystals, Corrosion Science 49 (2007) 2351-2361.

[20] Yoshioka H., Habazaki H., Kawashima A., Asami K., Hashimoto K., The Corrosion Behavior of Sputter-Deposited Al-Zr Alloys in 1-M HCl Solution Corrosion Science 33 (1992) 425-436.

[21] Shaw B.A., Davis G.D., Fritz, T.L. Rees B.J., Moshier W.C., The Influence of Tungsten Alloying Additions on The Passivity of Aluminum, J.Electrochem.Soc. 138 (1991) 3288-3295.

[22] Moshier W.C., Davis G.D., Ahearn J.S., and Hough H.F., Influence of Molybdenum on the Pitting Corrosion of Aluminum Films, J. Electrochem. Soc. 133 (1986) 1063-1064.
[23] Asami K., Habazaki H., Inoue A., Hashimoto K., New Frontiers Of Processing And Engineering In Advanced Materials Book Series: Materials Science Forum Editors: M.Naka, T.Yamane, Volume 502 (2005) 225-230.
[24] Yasuda M., Weinberg F., Tromans D., Pitting Corrosion of Al and Al-Cu Single-Crystals, J. Electrochem. Soc. 137 (1990) 3708-3715.
[25] Pride S.T., Scully J.R., Hudson J.L., Metastable Pitting of Aluminum and Criteria for the Transition to Stable Pit Growth, J. Electrochem. Soc. 141 (1994) 3028-3040.
[26] Moore K.L., Sykes J.M., Grant P.S., An Electrochemical Study of Repassivation of Aluminium Alloys with SEM Examination of the Pit Interiors Using Resin Replicas, Corrosion Science 50 (2008) 3233-3240.
[27] Trueba M., Trasatti S.P., Study of Al Alloy Corrosion in Neutral NaCl by the Pitting Scan Technique, Materials Chemistry and Physics 121 (2010) 523-533.
[28] Brunner J.G., May J., Höppel H.W., Göken M., Virtanen S., Localized corrosion of ultrafine-grained Al-Mg model alloys, Electrochimica Acta 55 (2010) 1966-1970.
[29] Newman R.C., Local Chemistry Considerations in the Tunneling Corrosion of Aluminum, Corrosion Science 37 (1995) 527-533.
[30] Füllmann T., Untersuchungen zum Lochfrass und zum Repassivierungsverhalten von Al und Al-Legierungen, Diploma Thesis, Erlangen 1993.

3

Pitting Corrosion Monitoring Using Acoustic Emission

A. Prateepasen
Acoustic Emission and Advanced Nondestructive Testing Center (ANDT), Department of Production Engineering, Faculty of Engineering, King Mongkut's University of Technology Thonburi, Bangmod, Toong-kru, Bangkok Thailand

1. Introduction

Damage from corrosion represents one of the most important problems in existing structures. Various types of traditional nondestructive testing (NDT) for example, Ultrasonic ThicknessMeasurement (UTM) or Magnetic Flux Leakage (MFL) are implemented to measure the growth of corrosion; each method, however, has its particular limitations. For the last two decades Acoustic Emission (AE), an advanced NDT method, has been used to monitor the severity of corrosion. Compared to the conventional NDT methods, it is less intrusive and has the advantage of real-time measurement. In this chapter, the development of pitting corrosion monitoring using AE is reviewed.

AE refers to the generation of transient elastic or stress waves during the rapid release of energy from localized sources within a material. The source of these emissions in metals is closely related to the dislocation movement accompanying plastic deformation and the initiation and extension of cracks in a structure under stress. AE systems are comprised of an AE sensor, an amplifier and filter, an acquisition and a data display. In the corrosion process, an AE parameter is extracted and then the relation with corrosion grading is determined. The parameters commonly used in corrosion applications include AE HIT, Even and AE energy.

The accurate forecast of corrosion level assists in maintenance planning. For instance, a storage tank or pipeline in the petroleum industry could fail as a result of pitting. In general, the severity of pitting corrosion is presented with two factors: the maximum pit depth and the pitting factor. A wide variety of methods were used to measure the severity of pitting corrosion. In this chapter, AE sources corresponded to the maximum pit depth are explained. The AE source generated from the pitting corrosion related to its mechanism is exposed. Experimental setup, results and discussions are briefly described. Research works in AE sources are deliberately reviewed. Three AE sources - hydrogen bubbles, breakage of passive film and pit growth - are proposed. The average frequency of the AE parameter was used to define the hydrogen bubbles. The oscillation, movement, and breakage of hydrogen bubbles generate acoustic stress waves in the same frequency band, which is related to the bubble diameter. An acoustic parameter, namely the 'duration time', can be used to classify the AE signal sources generated from the breakage of the passive film into two groups: the rupture of passive films and the pit growth signal. The results were explained based on the corrosion mechanism and the electrochemical analysis.

In practice, the location of the pitting can be calculated by using the difference in the arriving time of each pair or group of sensors. In this chapter, a novel source location system using an FPGA-PC system was utilized to calculate the arrival time difference from signals received from an array of three AE sensors on the specimen. The system consists of the AE sensors, pre-amplifiers, a signal conditioning unit, an FPGA module, a PC and a data acquisition. Experimental results and errors are shown.

2. Acoustic Emission testing (AE)

Nondestructive Testing can be defined as the development and application of methods to examine materials or components, in ways that do not impair future usefulness and serviceability, in order to detect, locate, measure and evaluate flaws; to assess integrity, properties and composition; and to measure geometric characteristics (American Society for Testing and Materials (ASTM E1316, 2001).

Acoustic emission (AE) is a powerful method for non-destructive testing and material evaluation. Older NDT techniques such as radiography, ultrasonic, and eddy current detect geometric discontinuities by beaming some form of energy into the structure under test. AE is different: it detects microscopic movement, not geometric discontinuities. AE, by definition, is a class of phenomena whereby transient stress/displacement waves are generated by the rapid release of energy from localized sources within a material, or the transient waves so generated (ASTM E1316, 2011). AE is the recommended term for general use. Other terms that have been used in AE literature include (1) stress wave emission, (2) microseismic activity, and (3) emission or acoustic emission or emission with other qualifying modifiers. These elastic waves can be detected by microphones or transducers attached to the surface of the specimen. AE techniques have been used in many applications such as in material degradation, leak and flow, solidification, and machining.

AE is a passive technique. The growing defect makes its own signal and the signal travels to the detecting sensors. The main benefits of AE compared to other NDT methods are that AE is a real-time method and it is less intrusive. The discontinuities of defects can be detected by AE at an early stage when they are occurring or growing. AE techniques can be used as a warning system before the testing material is severely damaged. AE requires access only at the sensors; while on the other hand, most other NDT techniques require access to all regions inspected.

In order to detect AE events, a transducer is required to convert the very small surface displacement to a voltage. Displacements as small as 10^{-14} metre [(Course Handbook for SNT-TC-1A (1991)] can be detected by the use of the most sensitive sensors. The most common type of transducer is piezoelectric, which is sensitive, easy to apply, and cheap. A couplant is needed for good transmission, and is usually achieved by grease or ultrasonic couplants, together with some means of applying force to maintain contact.

2.1 AE systems

An AE system consists of AE sensor, cable and pre-amplifier, and AE data-acquisition, as illustrated in Figure 1. There are two types of piezoelectric transducer: resonant transducers and broadband transducers. The principal or resonant frequency of a piezoelectric element depends on its thickness. The piezoelectric element is unbacked or undamped in a resonant transducer, but a broadband transducer has an element that is backed with an attenuating medium. The frequencies of most AE-resonant transducers lie in the range of 100 kHz to 1 MHz. However, it was found that the frequency of 30 kHz is often used for corrosion detection. Resonant sensors are more sensitive than broadband types because of the gain provided by mechanical resonance. Broadband sensors are used when the object of interest has the frequency spectrum of AE, but they do not have as high sensitivity as resonant transducers. Because of the reliance on mechanical resonance, resonant sensors can be used to detect preferentially a frequency range which has been shown from previous experience to give a good indication of the AE changes. Alternatively, a broadband sensor can be used and the required frequency is selected by filters.

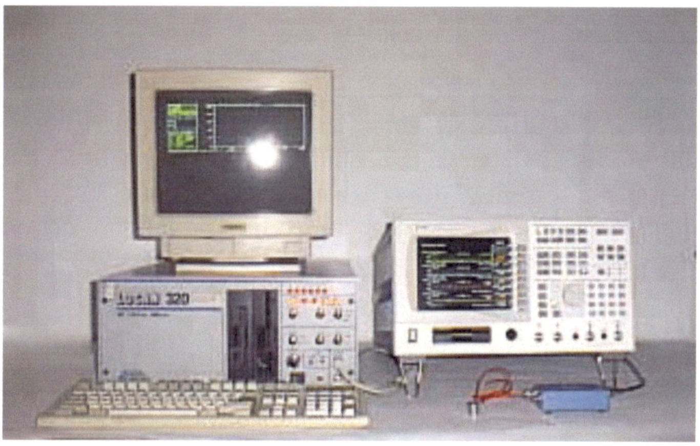

Fig. 1. Component of AE system (Prateepasen A., 2007)

Elastic waves emitted from materials can be divided into 2 types based on their appearance: burst and continuous. A burst emission is a signal, oscillatory in shape, whose oscillations have a rapid increase in amplitude from an initial reference level, generally that of the background noise, followed by a decrease, generally more gradual, to a value close to the initial level. Acoustic emission released from pitting corrosion is of the burst type. A continuous emission is a qualitative term applied to acoustic emission when the bursts or pulses are not discernible. (A pulse is an acoustic emission signal that has a rapid increase in amplitude to its maximum value, followed by an immediate return.)

2.2 AE waveform parameters

The parameters commonly used to predict severity in the corrosion process are AE ring down count, AE HIT, AE Even and AE energy or AErms. An AE burst, the typical AE signal from the corrosion process, is shown in Figure 2 and can be described by the parameters as follows.

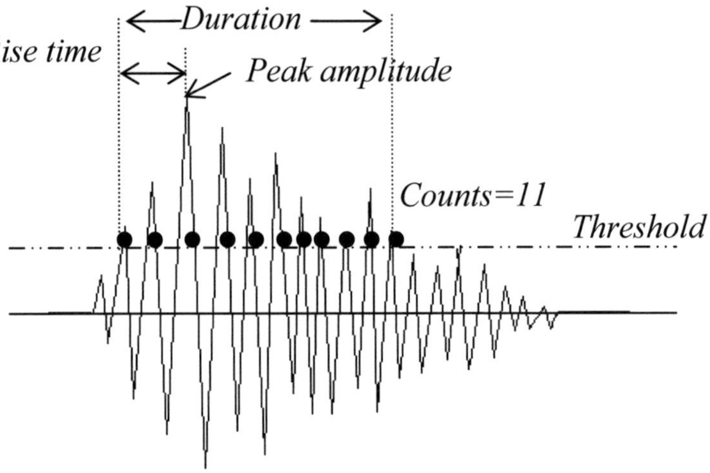

Fig. 2. Definition of AE waveform parameters

Ring down count is the number of times a signal exceeds a pre-set threshold. This is a simple measure of the signal size, since larger signals typically give more counts. Electronically, this is a very easy measurement, and it was the first to come into widespread use. By summing the counts from all detected emissions, one has a convenient measurement of the total emission from the specimen or structure. The number of counts (N) can be calculated by (Ronnie K.,1987)

$$N = \frac{\omega}{2\pi B} \ln \frac{V_0}{V_t} \tag{1}$$

where ω = angular frequency
B = decay constant (greater than 0)
V_0 = initial signal amplitude
V_t = threshold voltage of counter

AErms is the root mean squared value of the input signal. Since acoustic emission activity is attributed to the rapid release of energy in a material, the energy content of the acoustic emission signal can be related to this energy release. AErms can be defined as

$$V_{rms} = \left(\frac{1}{T}\int_0^T V^2(t)dt\right)^{\frac{1}{2}} \tag{2}$$

where V(t) = signal voltage function
T = period of time

AE HIT is the detection and measurement of an AE signal on a channel (ASTM 1316, 2011).

AE EVEN is an occurrence of a local material change or mechanical action resulting in acoustic emission.

AE ENERGY is the energy contained in an acoustic emission signal, which is evaluated as the integral of the volt squared function over time.

AE Duration time is the time between the point at which the event first exceeds the threshold and the point at which the event goes below the threshold. This parameter is closely related to the ring down count, but it is used more for discrimination than for the measurement of emission quantities. For example, long duration events (several milliseconds) in composites are a valuable indicator of delamination. Signals from electromagnetic interference typically have very short durations, so the duration parameter can be used to filter them out. In this chapter, duration time will be used to divide the AE source from pitting corrosion into two groups.

3. Corrosion

Corrosion is the disintegration of material into its constituent atoms due to chemical reactions with its surroundings. The type of corrosion mechanism and its rate of attack depend on the exact nature of the environment (air, soil, water, seawater) in which the corrosion takes place.

3.1 Corrosion process

The corrosion process is an ion transfer between anode and cathode poles which are on the material surface of one piece or a different piece depends upon the corrosion form and property of corroded material. The result of an electrochemical corrosion process on stainless steel is shown below. The corrosion process of metal comprises an oxidation and reduction reaction. At the anode pole, the oxidation reaction can be expressed by:

$$M \rightarrow M^{2+} + 2e-$$

and at the cathodes pole, the reduction reaction can be presented as:

$$2H+ \ + \ 2e- \rightarrow \ H2\ ; \ O2 \ + \ 4H+ \ + \ 4e- \rightarrow 2H_2O;\ M^{3+} + e^{-} \rightarrow M^{2+}$$

In the reaction, the hydrogen bubble is formed:

$$\left(Mn^{+} + nCl^{-}\right) + nH_2O \rightarrow M(OH)_n + nM^{+} + nCl^{-}$$

The chloride ion is absorbed on the material surface and breaks the passive film.

3.2 Acceleration of corrosion by electrochemical process

To study the relation between AE parameters and corrosion severity, pit growth is accelerated by the electrochemical process. Electrochemical polarization methods can be classified into two types: controlled potential (potentiostatic and, potentiodynamic) and controlled current (galvanostatic). The method used to accelerate pitting corrosion in the experiments in this chapter is potentiostatic and potentiodynamic. The applied polarizing current to control potential between working electrode (WE) and reference electrode (REF) at any prescribed value was adjusted by a potentiostat automatically. The diagram in Figure

3 shows the potentiostatic circuit, which consists of a potential and a current measuring element. The current (I) polarizes the WE to the prescribed potential with respect to REF, which remains at a constant potential with little or no current passing through the potential measuring circuit. The polarization curve from the potentiodynamic method allows the study in detail of the important parameters affecting formation and growth of passive films (Ecorr) and corrosive propagation (Epit).

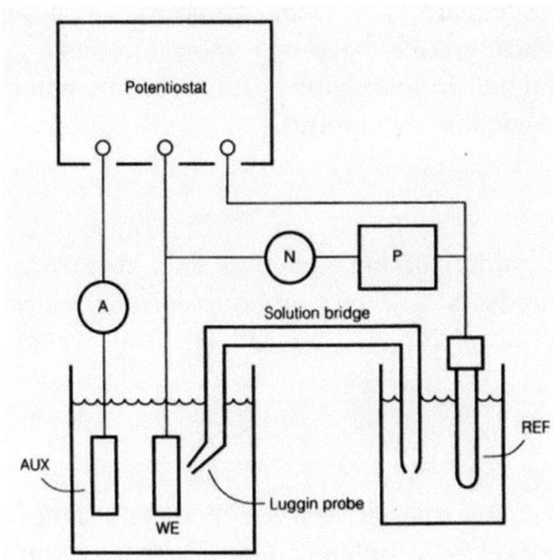

Fig. 3. Schematic circuitry diagram of potentiostatic and potentiodynamic methods

3.3 Monitoring and prediction of corrosion severity

The severity of pitting corrosion is measured by two factors: the maximum pit depth and the pitting factor. The pitting factor is the ratio of the depth of the deepest pit. At present, there are various methods to monitor corrosion (Jirarungsatian& Prateepasen, 2010) such as (1) failure records and visual inspection, (2) weight loss coupons, (3) spools and subs, (4) brine analyses, (5) deposit analyses, (6) three-electrode measurement technique, (7) electronic resistance instruments, (8) hydrogen patches and probes, (9) inhibitor residual analyses, (10) caliper surveys, and (11) A/C impedance and electrochemical noise instruments. Generally, the main aim of corrosion measurement is to find out the corrosion severity in order to forecast the test object's remaining life. However, each method has its limitation, such as the lack of real-time and difficulty in accessing all of the examination area. Nowadays, the Acoustic Emission method, which is a real-time NDT technique, is less intrusive and can inform of the severity of the corrosion.

3.4 Mechanism of pitting corrosion

In order to understand the AE source released from the pitting corrosion process, the mechanism of pitting corrosion will be explained. The elementary pitting process is divided into three steps as shown in Figure 4.

1. Pit nucleation: in this stage a small area of bare, un-filmed, passive surface of metal is formed.

2. The development of a metastable pit: either the formation of a stable pit or the repassivation has occured.
3. The growth of a stable pit: the metal is damaged and the pit is propagated in this stage.

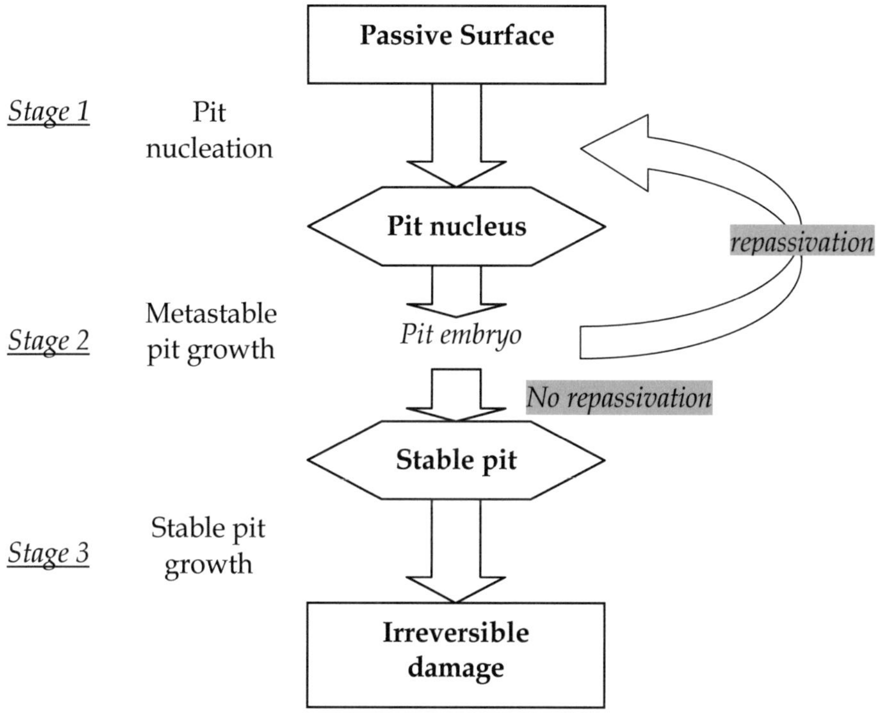

Fig. 4. Multistep mechanism for the onset of a stable pit (Jirarungsatian& Prateepasen, 2010)

4. Detection of pitting corrosion by AE

Researches in the area of pitting corrosion monitoring by AE have been the focus of several research groups. AE source generated from corrosion has been reported.

4.1 Literature review of the identification of AE source in corrosion process

A variety of corrosion types including uniform corrosion, pitting corrosion, crevice corrosion, stress corrosion cracking (Bosch, 2000; Cakir, 1999; Leal & Lopez, 1995; Xidong, 2005, Yoon et al., 2000), abrasion corrosion and erosion corrosion (Burstein & Sasaki, 2000; Ferrer & Labeeuw, 2000; Sasaki Oltra et al., 1995), has been studied and found to be correlate with AE. All of previous works presented the analogous conclusion that the AE technique can detect corrosion efficiently. However, the major disagreement is the corrosion mechanism that releases the acoustic wave. AE count number can be related with the corrosion rate in various corrosive conditions. In addition, the AE sources are corrosion activity and the hydrogen bubble occurrence on an electrode (Mansfield & Stocker, 1979). The frequency analysis of AE signals in the abrasion corrosion process was studied and it was concluded that the sources of AE are the impact of glass beads and gas bubbles (Ferrer, 1999).

Recently, researches in AE for source identification were proposed. Passive film breakage, bubble formation and the other actions in the corrosion process were considered as sources of the AE signal (Mirakowski, 2010). Experimental setup to eliminate AE source from gas bubble activity has been attempted in order to study the correlation between AE signal, which is AE count, and pitting corrosion in the electrochemical method (Mazille & Tronel, 1995; Mazille et al., 2001). The experiment used to study the acoustic emission generated by heating the metals and alloys confirmed that the AE signal was released during the phase transformation of specimens, and the AE parameter , 'count', was selected to explain the relation (Liptal, 1971). In the controlled potential of crevice corrosion monitoring, the research concluded that the AE signal in the monitoring was detected from all activity sources in the corrosion process including gas bubble formation (Kim & Santarini, 2003). The gas bubble activity also releases the acoustic wave. Therefore, the researches of the corrosion process which generated the gas bubble presented the identical discussion concluding that the AE source was bubble activity by comparison of AE counts with and without bubbles (Rettig & M. J. Felsen, 1976; Ferrer & Andrès, 2002). The AE event in an investigation of pitting corrosion using potentiodynamic methods expressed the relationship of the corrosion process with the AE source such as bubble activity (Darowicki, 2003). Numerous researchers have investigated the source of AE in the corrosion process. AE parameters were selected to correlate with the pitting process. In this chapter, the source of AE will be presented. The experiment was designed to find out the AE source and implemented to predict the pit growth.

4.2 Development of research work in AE for corrosion monitoring in ANDT

The Acoustic Emission and Advanced Nondestructive Testing Center (ANDT), King Monkut's University of Technology (KMUTT) has been studying the detection of corrosion by using AE since 2001. The first research work was to study and confirm that AE can detect corrosion. The relationship between AE and the mechanism of the corrosion damage was discovered (Jirarungsatean et al., 2002a). The accuracy of source location based on a general commonly used concept, namely the difference of arriving time interval, was studied (Jirarungsatean et al., 2002b). Accordingly, a low-cost system used to locate the corrosion location using FPGA was developed (Jomdecha et al., 2004, 2007). The classification of corrosion severity was subsequently the focus in ANDT's researches. Corrosion severity was evaluated by acoustic emission and a competent classification technique was implemented (Saenkhum et al., 2003). Consequently, a novel classification technique was developed to increase prediction performance (Prateepasen et al., 2006a). The effect of sulfuric acid concentration on AE signals obtained from uniform-corrosion was presented (Prateepasen et al., 2006b). Acoustic emission sources released from both uniform and pitting corrosion processes were investigated (Prateepasen et al., 2006c). More details of AE source recognition obtained from pitting corrosion will be mentioned in the next section.

4.3 Experimental work to validate the detection of corrosion by AE

The first step of the ANDT research group in the area of corrosion was to show that AE could detect pitting corrosion. High-concentration acid was used to produce pitting corrosion (Jirarungsatean et al., 2002a). The experimental work was developed by using electrochemical control (Prateepasen et al., 2006c). A specimen made of stainless steel

(SS304) was subject to accelerated pitting by electrochemical control equipment as shown in Figure 5. The surface of the specimen was ground with 1200 grit silicon carbide paper, rinsed with distilled water, and dried in cool air. The electrochemical environment was 3% NaCl solution mixed with HCl to control its pH of 2, and the electrochemically applied constant current was controlled with a Solartron 1284. In this experiment, the specimen acted as a working electrode, a platinum mesh worked as the counter-electrode, and a Ag/AgCl (sat. KCl) worked as a reference. AE sensor R15 (PAC) was mounted at the surface of the specimen. The crystal's resonance frequency was 150 kHz. An AE signal captured by the sensor was fed to a preamplifier for 60dB gain. A wide-band filter was embedded in the preamplifier. Data acquisition used to record and analyze the AE signal was done by LOCAN 320. The frequency waveform was recorded by a spectrum analyzer (HP 89410A).

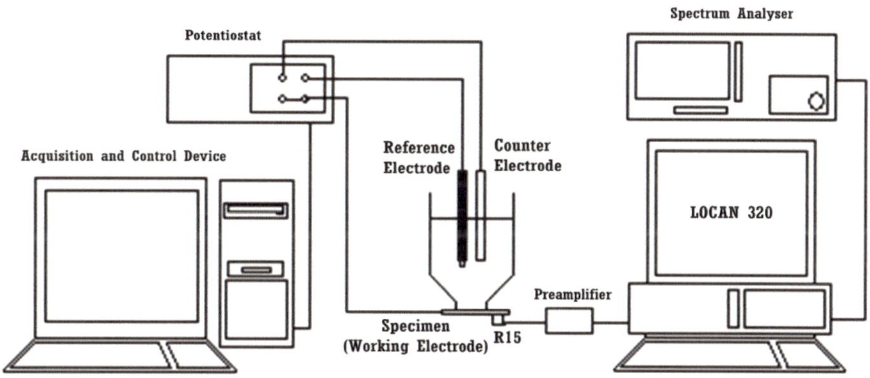

Fig. 5. Set up of AE system and electrochemical control equipment (Prateepasen et al., 2006c (pH of 2))

The pitting corrosion rate was controlled by the potentiostat that applied a constant potential on the specimen for twenty hours. Data received from the sample of SS304 was analyzed. The average frequency and time waveforms were collected every ten minutes. The relationship between the time and voltage was recorded by the potentiostat.

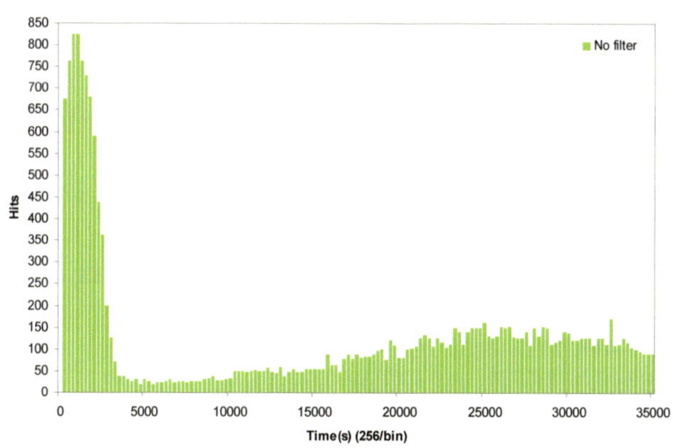

Fig. 6. Hit parameters in time domain (Prateepasen et al., 2006c)

The Epit value was identified from the polarization curve of potentiodynamic tests. The electrical noise was detected by the AE sensor. However, it was a small signal and was filtered out by the threshold setting. The frequency response of the AE in this experiment is illustrated in Figure 6 where the average frequency of corrosion signals in frequency domain analysis is around 110 kHz. The time domain analysis was shown. AE parameter hits and amplitude showed correlation with pit growth. There were high numbers of hits in the initial corrosion period; then the number of hits decreased, and rose again in the next stage of pitting. Figure 7 shows a pitting corrosion in the specimen.

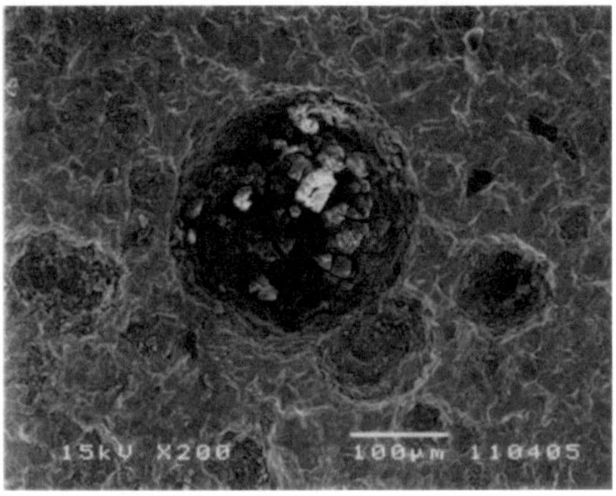

Fig. 7. Pitting corrosion in the specimen

5. AE source recognition in pitting corrosion process using acoustic parameter

Continuing researches have been performed, and AE source recognition in the pitting corrosion process was proposed by Jirarungsatian & Prateepasen (2010). The AE source was assumed to divide into three groups. The mechanism of pitting corrosion shown in Figure 4 and the theory of cut-off frequency of the bubble activity including the results from the experiment were used to support the assumption. Three AE sources are as follows:

1. Bubble activities by using the resonance frequency (formation oscillation and collapse).
2. Primary passive film and repassive film breakage by using duration time relating to the initial time of metastable pit formation according to electrochemical analysis.
3. Pit growth or pit propagation related to the time of occurrence and pitting corrosion mechanisms.

Experimental work has been done to validate the cut-off frequency. The electrochemical potential was set up and kept constant with the potential level of 0.3041 V (Ag/KCl) using the potentiostat. Its level was chosen from a potentiodynamic test. The specimen, austenitic stainless steel (SS304) sized $4\times6\times0.05$ cm^3, was used as the working electrode. The test was set in two conditions; one, the AE signal with bubble signal, and the other, a separate bubble signal.

Experiment setup in Figure 8 shows that the AE sensor was mounted under the container located near the hydrogen bubble. In this case, AE breakage of the hydrogen can be captured by the sensor.

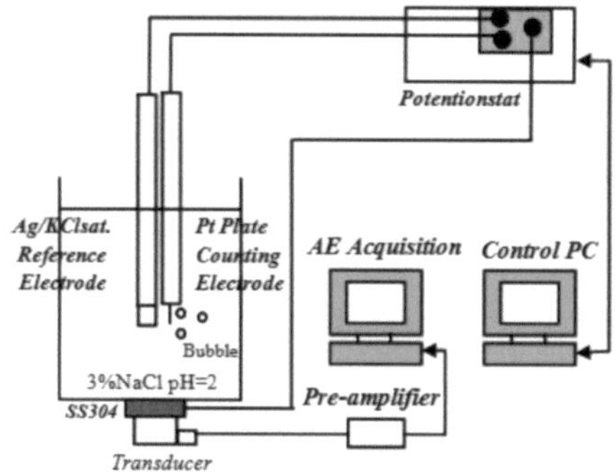

Fig. 8. Experimental setup for corrosion analysis *with bubbles* by AE (Prateepasen & Jirarungsatian, 2011, (pH=2,potential level =0.3041V))

The experiment was set up to eliminate the signal released from bubbles. The counting electrode was moved away from the AE sensor by using an Electrolyte Bridge (shown in Figure 9).

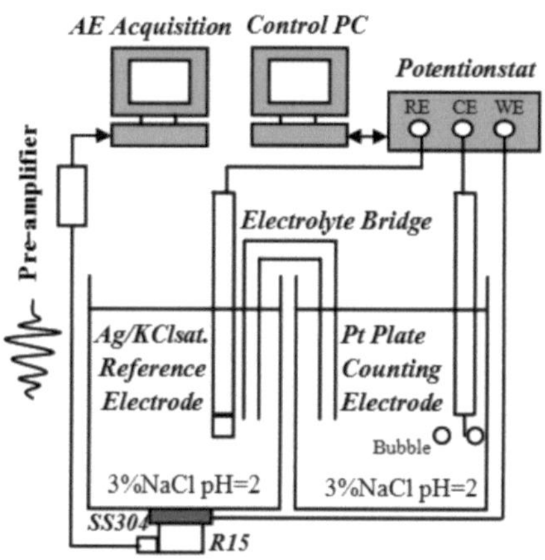

Fig. 9. Experimental setup of corrosion analysis *without bubble signals* by AE (Prateepasen & Jirarungsatian,2011, (pH=2,potential level =0.3041V))

In the theory of cutoff frequency of the bubble, the cutoff frequency of the bubble breakage can be calculated by the following equation (Leighton, 1994).

$$\nu_0 \approx \frac{1}{2\pi D}\sqrt{\frac{3\gamma P_0}{\rho_0}} \tag{3}$$

From the experimental results, the largest bubble diameter obtained from pitting corrosion was approximately 0.9 millimeters (stable bubble size, D). Typical bubbles are shown in Figure 10. The value of the liquid density (ρ_0) was 1000 kg/m^3. The bubbles were assumed to be filled with an ideal gas, with a specific heat coefficient γ = 1.4, and the liquid was assumed to be nearly incompressible with static pressure P_0 = 3.110×107 kg/m.s^2.

The cut-off frequency calculated from equation 3 is 125 kHz. It is shown that the AE activity above 125 kHz is released from bubble activity.

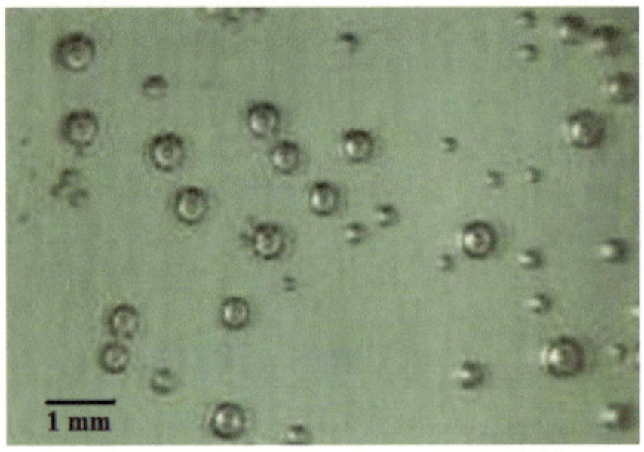

Fig. 10. Examples of hydrogen bubbles on the test specimen in a pitting corrosion process

Results from the experiment showed that the frequency of most AE activities in the time domain collected from the setup of AE with bubble is above 125 kHz. In addition, AE data captured from the setup of the separated bubble exhibited the frequency below 125 kHz. The AE signal of all three sources (with bubble breakage) in the time domain was plotted. A scatter plot of AE duration time and average frequency is shown in Figure 11. Figure 11b shows the AE signal obtained from the bubble breakage. The calculated cut-off frequency at 125 kHz was used to separate the bubble source from all AE sources in Figure 11a.

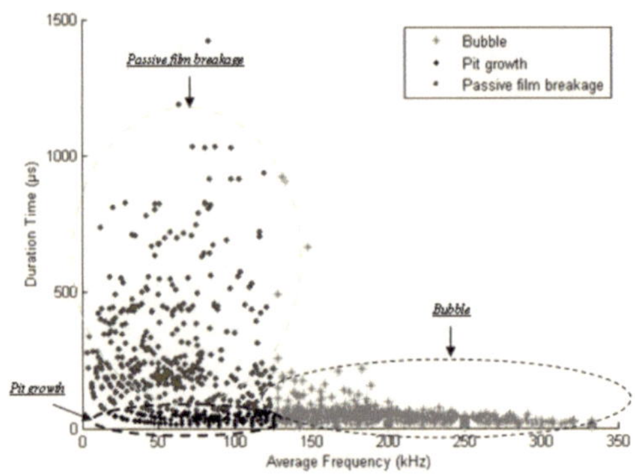

a) All AE sources (Jirarungsatian& Prateepasen, 2010)

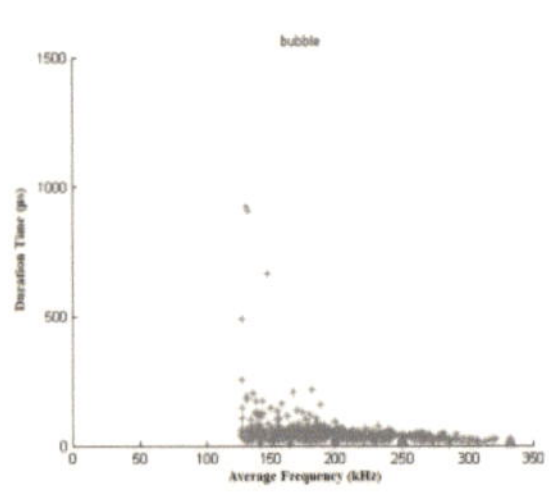

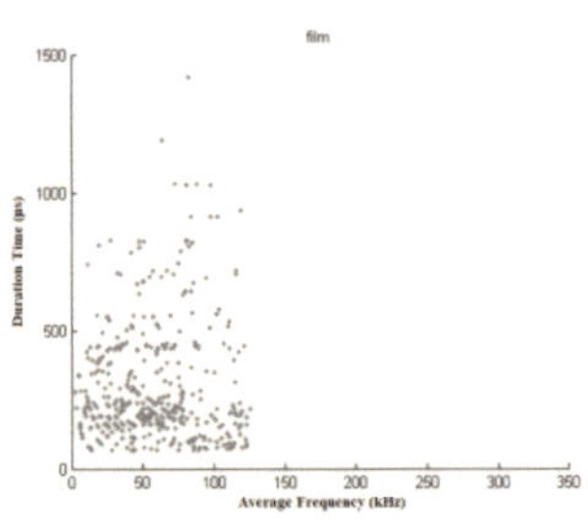

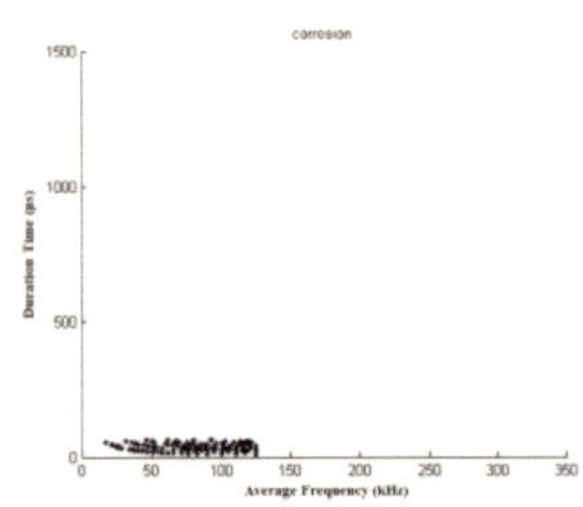

b) Bubble (*) c) Passive film rupture (♦) d) Pitting corrosion (♦)

Fig. 11. Source identification of pitting corrosion by AE parameter (Jirarungsatian& Prateepasen, 2010)

In order to classify the AE signal sources obtained from pit growth and passive film breakage, AE duration time of 65 microseconds was used as a threshold. AE released from breakage of passive film decreased at the time of the initial metastable pit formation. In this stage, re-passivating took place, and then the amount of film rupture was reduced (Jirarungsatian& Prateepasen, 2010). The AE signals obtained from the pit growth or propagation had an average duration of below 65 microseconds and frequency between 3-125 kHz. The cumulative total of detected signals (cumulative hits) of AE from pit growth was lower than that from film rupture during the pit nucleation stage.

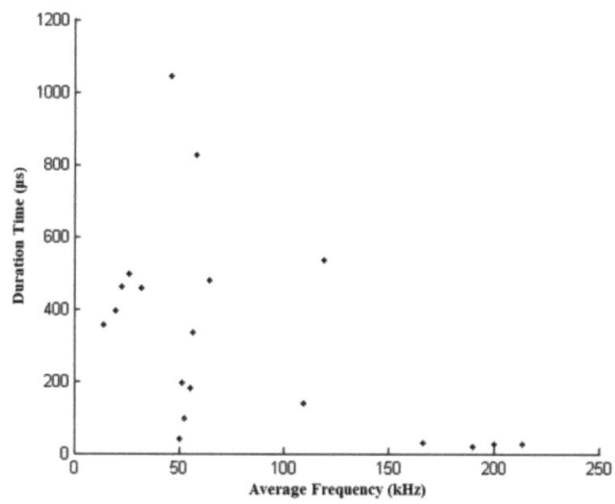

Fig. 12. AE signal from pitting corrosion tests of an undetectable acoustic bubble (Jirarungsatian& Prateepasen, 2010)

6. Novel work in acoustic emission for pitting monitoring

The development of pitting corrosion monitoring will be presented in three parts: implantation of AE source recognition, classification of severity, and location prediction by using FPGA.

6.1 Implementation of acoustic emission source recognition for corrosion severity prediction

Knowledge of source recognition for corrosion severity was implemented by Prateepasen & Jirarungsatian, (2011). To show the relationship between each AE source and the corrosion rate, the AE hit rate in the time domain was used to predict the corrosion rate. The total AE sources of pitting corrosion were divided into two groups: bubble breakage and metal corrosion including rupture of a passive film. Consequently, each AE source was plotted to show the relationship between the hit rate and the severity of pitting corrosion, represented by the pit depth (Figure 13). In this research, the severity of corrosion varied from 0.1 to 0.5 mm of pitting.

Both the progression of pitting with rupture of the passive film, and bubble breakage alone, showed a high correlation with the pit depth. In the other words, each AE source can be used to predict the corrosion rate. The benefit of source recognition can be utilized in the corrosion field test. The environmental noise can be avoided by selecting the AE source at a frequency range that is different from the background noise.

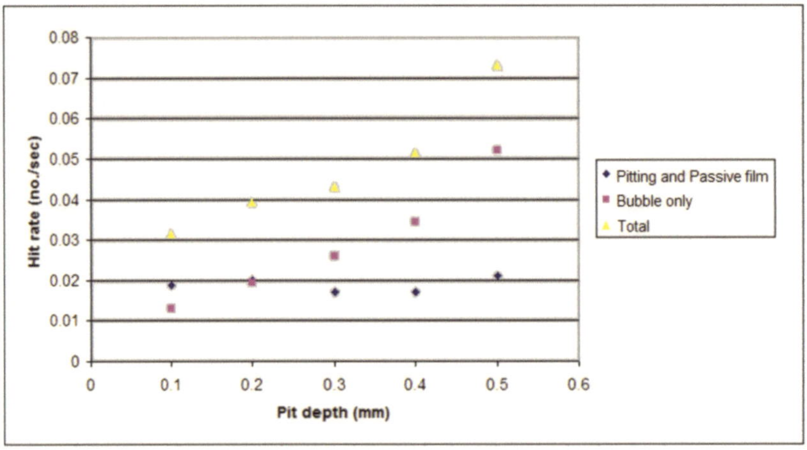

Fig. 13. The relationship between the AE hit rate and pit depth of various AE sources Prateepasen & Jirarungsatian, (2011)

6.2 Classification of corrosion detected by AE signal

The corrosion severity is ranked roughly into five levels based on the depth of corrosion by using a Feed-Forward Neural Network (Saenkhum et al., 2003) and Probabilistic Neural Network (Jirarungsatean et al., 2003]. The classification performance is very good. The error prediction rate is low. After that, the practical classification techniques based on Bayesian Statistical Decision Theory, namely Maximum A Posteriori (MAP) and Maximum Likelihood (ML) classifiers, were presented. A mixture of Gaussian distributions is used as the class-conditional probability density function for the classifiers. Although the mixture model has several appealing properties, it still suffers from model-order-selection and initialization problems. A semi-parametric scheme for learning the mixture model is therefore introduced to solve the difficulties. The result of its performance was compared with a Conventional Feed-Forward Neural Network and Probabilistic Neural Network. The results show that our proposed methods gave much lower classification-error rates and also far smaller variances than those of the classifiers.

Classifier	Overall Performance			
	Minimum	Maximum	Average	Variance
MAP	98.97	99.74	99.26	0.010
ML	98.68	99.52	99.03	0.014
FFNN (Saenkhum et al., 2003)	84.73	98.85	90.65	3.115
PNN (Jirarungsatean et al., 2003]	89.62	99.13	97.33	1.418

Table 1. Overall Performance of Classifiers

6.3 FPGA-PC AE low-cost system for source location

The AE system generally can inform of the location of the AE source. The technique uses the measurement of time differences for acceptation of the stress wave at a number of sensors in an array. There are three types of location techniques, which are linear location, two dimensions, and three dimensions. In the application of corrosion, the two-dimensional technique is generally used.

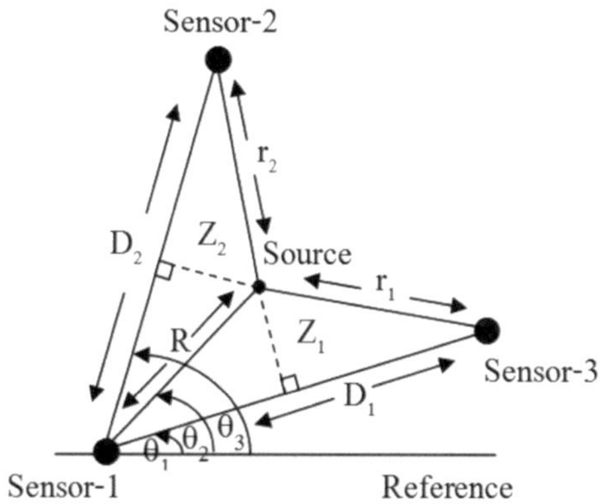

Fig. 14. Three-sensor array in two dimensions (Jomdecha et al., 2007).

In Figure 14, the distance and angle of the AE source from sensor 1 can be determined by solving the following set of equations (McIntire, 1987) (Jomdecha et al., 2007).

$$R = \frac{1}{2}\frac{D_2^{\,2} - \Delta t_2^{\,2}V^2}{\Delta t_2 V + D_2\cos(\theta_3 - \theta_0)}, \text{ and } R = \frac{1}{2}\frac{D_1^{\,2} - \Delta t_1^{\,2}V^2}{\Delta t_1 V + D_1\cos(\theta_0 - \theta_1)} \tag{4}$$

where Δt_1 and Δt_2 are the time differences between sensors 1 & 2 and 1 & 3, respectively.

The time difference parameters can be obtained by simple thresholding or cross-correlation techniques. Solving the set of equations gives the location of the source in a polar form (R, θ) as illustrated in Figure 14.

A novel low-cost location system, namely FPGA, was proposed by Jomdecha et al. (2004, 2007). It was designed for multi-channel high speed counters with serial communication using VHDL (Very High Speed Integrated Circuit Hardware Description Language). When all the AE sensors were triggered, the counting data from the FPGA-based electronic front-end was sent to the PC via an RS-232 port. The FPGA was operated at approximately 30 ms per cycle. The PC was employed for analysis of the arrival time differences of sensors and, finally, the coordinates of corrosion could be estimated.

In order to prove the accuracy of identifying the location of pitting corrosion, an experiment was performed. The electrochemical environment was a 3% NaCl solution with pH 2 to facilitate the pitting corrosion mechanism. The pitting potential (E_{pit}) was controlled electrochemically with a potentiostat (Solartron 1284) to accelerate pitting corrosion. The arrival time difference of the AE signal received from the array of three AE sensors on the specimen was determined by the FPGA-PC system. The system consisted of the AE sensors, pre-amplifiers, signal conditioning unit, FPGA module, PC, and LOCAN320 AE analyzer. A two-dimensional or planar representation was used to demonstrate the results of corrosion location. Figure 15 shows a diagram of the experimental setup for corrosion source localization.

Figure 16 shows the comparison of pitting location between the real location and the location detected by FPGA. It was shown that the locations of pitting and crevice corrosion were spread around the region of the corrosion sources.

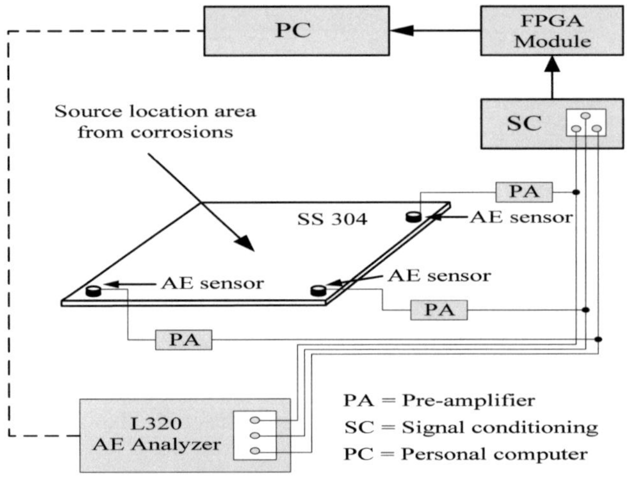

Fig. 15. Three AE sensor array and the system for source location (Jomdecha et al., 2007)

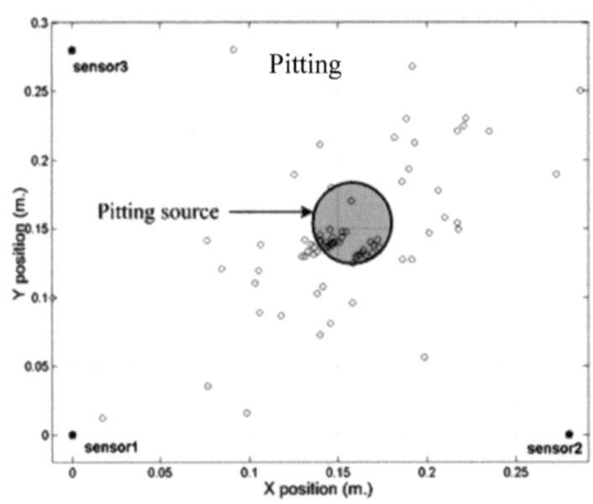

Fig. 16. Comparison of pitting location between real location and location detected by FPGA (Jomdecha et al., 2007)

7. Conclusions

Pitting corrosion detected by AE was explained. AE sources released from the pitting corrosion process were studied to determine the mechanism, and explained. Experiments were performed to classify types of AE source into three groups: bubble activities, primary passive film and repassive film breakage, and pit growth or pit propagation. Subsequently, each type was used to plot the relation with pitting growth. It is a novel technique and provides important benefit to implement in corrosion field test. An artificial network was utilized to classify pitting severity. A neural network was first used and then a SEMI PARAMETRIC technique was developed. The location of pitting was shown by the novel low-cost technique, FPGA. The advantage of AE compared to other techniques is that AE is a real-time and less intrusive technique.

8. References

ASTM E1316American Society for Testing and Materials, 2001, Standard Terminology for Nondestructive Examinations

Burstein, G.T., Sasaki, K., Effect of impact angle on the slurry erosion - corrosion of17 304L stainless steel, Wear, 240(1-2) May (2000) 80-94.

Cakir, A., Tuncell, S., Aydin, A, AE response of 316L SS during SSR test underpotentiostatic control, Corrosion Science, 41(6) (1999) 1175-1183.

Darowicki, K., Mirakowski, A., Krakowiak, S., Investigation of pitting corrosion of stainless steel by means of acoustic emission and potentiodynamic methods, Corrosion Science, 45(8) (2003) 1747-1756.

Ferrer, F., Idrissi, H., Mazille, H., Fleischmann, P., Labeeuw, P., On the potential oflacoustic emission for the characterization and understanding of mechanicalDamaging during abrasion - corrosion processes, Wear, 231(1) June (1999) 108-115.

Ferrer, F., Idrissi, H., Mazille, H., Fleischmann, P., Labeeuw, P., A study ofabrasion - corrosion of AISI 304L austenitic stainless steel in saline solution usingacoustic emission technique. NDT & E International, 33(6) September (2000)363-371.

Jirarungsatean, C., Prateepasen, A., and Thungsuk, P., "Corrosion monitoring using Acoustic Emission", pp.152-154, Science & Material Technology, Thailand, Vol.2, 2002a.

Jirarungsatean, C., Prateepasen, A., and Thungsuk, P., "The corrosion locating using Acoustic Emission Technique", pp. 376-387, IE network Conference 2002, October, Thailand, 2002b.

Jirarungsatean, C., Prateepasen, A., and Kaewtrakulpong, P., 2003, "*Pitting Corrosion Monitoring of Stainless Steels by Acoustic Emission*", Corrosion Control and NDT, November 23-26, Melbourne, Australia.

Jirarungsatian, C. and Prateepasen, A., 2010, "Pitting and Uniform Corrosion Source Recognition using Acoustic Emission Parameters", Corrosion Science, Vol. 52. No. 1, January, pp. 187-197.

Jomdecha, C., Prateepasen, A., Kaewtrakulpong, P., and Thungsuk, P., 2004, "Corrosion-Source Location by an FPGA-PC Based Acoustic-Emission System", IEEE TENCON 2004, November, 21-24, Lotus Hotel Pang Suan Kaew, Chiangmai, Thailand.

Jomdecha, C., Prateepasen, A., and Kaewtrakulpong, P., 2007, "A Study on Sorce Location Using an Acoustic Emission System on Various Types of Corrosion in Industry", NDT&E International, Vol. 40, No. 8, December, pp. 584-593.

Kim, Y.P., and Santarini, G.,NDT & E International, Ability of acoustic emission technique for detection and monitoring of crevice corrosion on 304L austenitic stainless steel, Vol. 36, December 2003, pp. 553-562.

Leighton, T.G., 1994, The Acoustic Bubble, London, San Dieg

Mansfield, F., Stocker, P.J., Acoustic Emission from Corroding Electrodes. Corrosion Engineers, 35(12) (1979) 541-544.

Mazille, H., Rothea, R., Tronel, C., An acoustic emission technique for monitoringpitting corrosion of austenitic stainless steels, Corrosion Science, 37(9) (1995)1365-1375.

Mirakowski, A.,: http://www.korozja.pl/1_01_02.pdf.

Oltra, R., Chapey B., Renaud L., Abrasion-corrosion studies of passive stainlesssteels in acidic media: combination of acoustic emission and electrochemicaltechniques, Wear, 186-187 August (1995) 533-541.

Prateepasen, A., Kaewtrakulpong, P. and Jirarungsatean, C., 2006a, "Semi-Parametric Learning for Classification of Pitting Corrosion Detected by Acoustic Emission", Journal of Key Engineering Materials, Vols. 321-323, pp. 549 – 552.

Prateepasen, A., Jirarungsatean, C. and Tuengsook, P., 2006b, "Effect of Sulfuric Acid Concentration on Acoustic Emission Signals Obtained from Uniform-Corrosion", Journal of Key Engineering Materials, Vols. 321-323, pp. 553 – 556.

Prateepasen, A., Jirarungsatean, C. and Tuengsook, P., 2006c, "Identification of AE Source in Corrosion Process", Journal of Key Engineering Materials, Vols. 321-323, pp. 545 – 548.

Prateepasen, A. (December 2007), Non-destructive testing in welds and researches, Chulalongkorn University Press, ISBN 978-974-456-679-9, Bangkok, Thailand.

Prateepasen, A. and Jirarungsatian, C. , May 2011, Implementation of acoustic emission source recognition for corrosion severity prediction. Corrosion 67

Rettig, T.W. , and Felsen, M.J., Corrosion-NACE, Vol. 32, No. 4, 1976Acoustic emission for detection of corrosion under insulationFerrer F.; Faure T.; Goudiakas J.; Andres E, ,Acoustic emission study of active-passive transitions during carbon steel erosion-corrosion in concentrated sulfuric acid , Corrosion Science, Vol. 44, Issue 7, July 2002, pp. 1529-1540

Ronnie, K.M. and Paul, M.,1987, Nondestructive Testing Handbook volume 5 Acoustic Emission, American Society for Nondestructive Testing,ISBN 0-931403-02-2, USA.

Saenkhum, N., Prateepasen A. and Kaewtrakulpong P., "Classification of Corrosion Detected by AE signal", IMECE'2003, Washington, D.C., Nov., 2003

Xidong, F. Ju, C. Lin, Research on the brittle fracture of FRP rods and itsacoustic emission detection, Power Engineering Society Winter Meeting, IEEE2000, 4(23-27) January (2000) 2812 – 2816.

Yoon, D.J., Weiss W.J., Shah S.P.,2000 Detecting the extent of corrosion with acoustic19 emission, Transportation Research Record, (1698) 54-60.

4

Oscillatory Phenomena as a Probe to Study Pitting Corrosion of Iron in Halide-Containing Sulfuric Acid Solutions

Dimitra Sazou*, Maria Pavlidou,
Aggeliki Diamantopoulou and Michael Pagitsas**
Department of Chemistry, Aristotle University of Thessaloniki, Thessaloniki
Greece

1. Introduction

Oscillatory phenomena and other nonlinear phenomena such as bistability and spatiotemporal patterns are frequently observed in metal and alloy electrodissolution-passivation processes (Hudson & Tsotsis, 1994; Koper, 1996a; Krischer, 1999; Krischer, 2003b). Current oscillations during the Fe electrodissolution-passivation in acid solutions were reported as early as 1828 (Fechner, 1828). Since then, metal|electrolyte interfacial systems have received considerable interest over the past three decades for many reasons. Among them, is that progress in the theory of nonlinear dynamical systems, achieved in parallel over last decades, has led to the formulation of new theoretical concepts and tools that could apply to electrochemical oscillators. Therefore, an understanding of the fundamental principles underlying the nonlinear phenomena observed in electrochemical processes has been considerably improved (Berthier et al., 2004; Eiswirth et al., 1992; Karantonis & Pagitsas, 1997; Karantonis et al., 2005; Karantonis et al., 2000; Kiss & Hudson, 2003; Kiss et al., 2003; Kiss et al., 2006; Krischer, 2003b; Parmananda et al., 1999; Parmananda et al., 2000; Sazou et al., 1993a). On the other hand, electrochemical systems can be readily controlled through the variation of the potential (under current-controlled conditions) or the current (under potential-controlled conditions) and have served as experimental model systems to implement and test new theoretical concepts.

Technological applications of nonlinear electrochemical phenomena in materials science are exemplified by their impact on electrodissolution, electrodeposition and electrocatalytic reactions (Ertl, 1998; Ertl, 2008; Nakanishi et al., 2005; Orlik, 2009; Saitou & Fukuoka, 2004). Another promising application might be the preparation of self-organized nanostructures such as TIO_2 nanotubes (Taveira et al., 2006). Regarding electrodissolution processes, the progress in defining the conditions for metal stability and dynamical transitions in metal|electrolyte systems is of fundament importance for metal performance and safety in natural environments (Hudson & Basset, 1991; Lev et al., 1988; Sazou et al., 2000b; Sazou et

---

* Corresponding author
** Deceased 26 April 2009.

al., 1993a; Sazou & Pagitsas, 2003b). The existence of passivity on metals is well recognized as the most important factor for the metal safe use in our metal-based civilization (Sato, 1990; Schmuki, 2002; Schultze & Lohrenger, 2000). It has been shown that depassivating factors, resulting either in uniform dissolution of passive films (general corrosion) or localized breakdown of an otherwise stable passivity on metals (pitting corrosion) give rise to temporal as well as spatiotemporal instabilities (Green & Hudson, 2001; Otterstedt et al., 1996; Sazou & Pagitsas, 2003b; Sazou et al., 2009; Wang et al., 2004). For Fe, these instabilities are more pronounced in pitting corrosion occurring in different corrosive media, but mostly, in those containing halides such as chlorides, bromides and iodides. This will be the main theme of this brief review.

In practice, chlorides are of a major concern due to their abundance in environments encountered in industry and in domestic, commercial and marine industry. Extensive studies have been carried out over the last century and continue aiming to estimate the conditions leading to local breakdown, gain a deeper understanding of mechanisms and processes underlying pitting and develop effective strategies of metal protection (Bohni, 1987; Frankel, 1998; Kaesche, 1986; Macdonald, 1992; Sato, 1982; Sato, 1989; Sato, 1990; Strehblow, 1995). Exploring the nonlinear dynamical phenomena associated with pitting corrosion of Fe and other metals might provide a new approach in investigating passivity breakdown from both mechanistic and kinetic points of view. Especially, electrochemical measurements and nonlinear dynamics in conjunction with new surface analytical techniques constitute a promising way towards studying pitting corrosion (Maurice & Marcus, 2006; Wang et al., 2004).

The onset of current and potential oscillations is the most common nonlinear behavior of the Fe | electrolyte system in acidic solutions containing halides, X^- ($X^- \equiv Cl^-, Br^-, I^-$) (Georgolios & Sazou, 1998; Koutsaftis et al., 2007; Li et al., 2005; Ma et al., 2003; Pagitsas & Sazou, 1999; Sazou et al., 2000a; Sazou et al., 2000b; Sazou et al., 1993b; Sazou et al., 1992). In particular, it was shown that adding small amounts of X^- in the Fe | n M H_2SO_4 system induces complex periodic and aperiodic current oscillations under potential-controlled conditions (Georgolios & Sazou, 1998; Koutsaftis et al., 2007; Li et al., 2005; Pagitsas & Sazou, 1999; Sazou et al., 2000a; Sazou et al., 2000b). These oscillations are of large amplitude and represent passive-active events emerged out of an extensive potential region. A gradual increase of halide concentration results in the establishment of a limiting current region out of the Fe passive state. This new state of Fe is accompanied by complex aperiodic current oscillations of small amplitude. The latter oscillations occur under mass-transport controlled conditions, which are established inside pits due to the formation of ferrous salt layers (Sazou & Pagitsas, 2003a; Sazou & Pagitsas, 2006a). Moreover, as was mentioned above, halides induce also potential oscillations under current-controlled conditions, associated with either early (Postlethwaite & Kell, 1972; Rius & Lizarbe, 1962; Sazou et al., 2009) or late stages (Li & Nobe, 1993; Li et al., 1993; Li et al., 1990; Strehblow & Wenners, 1977) of pitting corrosion.

Unambiguously, both current and potential oscillations include valuable information related to the kinetics of the oxide growth and its breakdown (Pagitsas et al., 2001; Pagitsas et al., 2002; Sazou & Pagitsas, 2003a; Sazou & Pagitsas, 2006a). Though, several investigations aiming to reveal and use profitably this information have been brought about some progress in understanding underlying processes, many aspects remain to be revealed and exploited.

In this article only few of these aspects will be touched upon. Emphasis is placed on displaying certain features of the oscillatory response of the halide-containing Fe | *n* M H_2SO_4 system that might be employed in establishing new tools, useful in detecting and characterizing the extent of pitting corrosion on passive Fe surfaces.

The chapter starts with a short section (section 2) in which basic information about measurements used in all sections is provided.

In section 3, the basics regarding the origin of oscillations in metal | electrolyte systems under corrosion conditions are discussed briefly to show that the halide-free Fe | *n* M H_2SO_4 system, being an N-NDR oscillator, is unlikely to display either current oscillations within the stable passive state, extended beyond the Flade potential, or potential oscillations under galvanostatic conditions. Thus oscillatory phenomena discussed in this chapter are the result of the interplay between pitting corrosion and basic dynamics of an N-NDR system that is transformed rather to an HN-NDR oscillator.

Section 4, displays briefly the characteristic current-potential, $I = f(E)$ or potential-current, $E = f(I)$ polarization curves of the halide-free Fe | 0.75 M H_2SO_4 system traced under potentiodynamic and galvanodynamic conditions, respectively. This section aims in demonstrating that in the absence of aggressive ions only single periodic oscillations occur. The fundamental physico-electrochemical processes underlying the mechanism of single periodic oscillations are briefly summarized.

Section 5, focuses on the nonlinear dynamics of the halide-perturbed Fe | 0.75 M H_2SO_4 system at relatively low halide concentrations. Halide-induced changes in $I = f(E)$ and $E = f(I)$ polarization curves are pointed out. By choosing appropriate potential and current values from $I = f(E)$ and $E = f(I)$ curves, current and potential time-series are traced under potentiostatic and galvanostatic conditions, respectively. Experimental results are analyzed in order to establish appropriate kinetic quantities as a function of either the potential or current and the halide concentration. Emphasis is placed here on how these quantities can be used in studying initiation of pitting at early stages.

Section 6, provides selected experimental examples displaying the nonlinear response of the halide-containing Fe | 0.75 M H_2SO_4 system at relatively high halide concentrations. It is thus concerned with late stages of pitting corrosion, which are exemplified by a different type of oscillations.

Section 7, includes an overview of conditions under which the nonlinear response of the halide-containing Fe | 0.75 M H_2SO_4 system appears and a summary of proposed diagnostic criteria, appropriate for characterization of pit initiation and its growth.

In section 8, conclusions are presented, while section 9 contains references.

2. Experimental

Electrochemical measurements were carried out using a VoltaLab 40 electrochemical system and the VoltaMaster 4 software from Radiometer Analytical. Additionally, a Wenking POS 73 potentioscan from Bank Elektronik was employed. It was interfaced with a computer, which was equipped with an analog-to-digital, and vice versa, converter PCL-812PG enhanced multi-Lab. Card (Advantech Co. Ltd). The maximum sampling rate of the PCL-

812PG card was equal to 30 kHz. The working electrode (WE) was the cross section of an iron wire with a diameter equal to 3 mm from Johnson Matthey Chemicals (99.9%) embedded in a 1 cm diameter PTFE cylinder (surface area=0.0709 cm^2). A volume of 150 ml was maintained in a three-electrode electrochemical cell. A Pt sheet (2.5 cm^2) and a saturated calomel electrode (SCE) were used as the counter (CE) and reference electrodes (RE), respectively. The Fe-disc surface was polished by wet sand papers of different grit size (100, 180, 320, 500, 800, 1000, 1200 and superfine) and cleaned with twice-distilled water in an ultrasonic bath. Solutions were prepared with H_2SO_4 (Merck, pro-analysis 96% w/w) and NaF or NaCl or NaBr or NaI, all from Fluka (puriss p.a.), using twice-distilled water. Measurements were carried out at 298 K, while N_2 was passed above the solution during the course of the experiment. A scanning electron microscope (SEM) JEOL JSM-840A was used for the Fe surface observation. Further experimental details can be found in previous studies (Pagitsas et al., 2003; Sazou & Pagitsas, 2003b; Sazou et al., 2009).

3. Origin of oscillations in corroding metal|electrolyte interfacial systems

Spontaneous oscillatory phenomena observed in metal electrodissolution-passivation reactions are often associated with multisteady-state current-potential (*I-E*) curves due to the occurrence of a region with negative differential resistance (NDR). NDR appears either in N-type (the electrode potential acts as activator, positive feedback variable) (Fig. 1) or S-type (the electrode potential acts as inhibitor, negative feedback variable) *I-E* curves (Fig. 2).

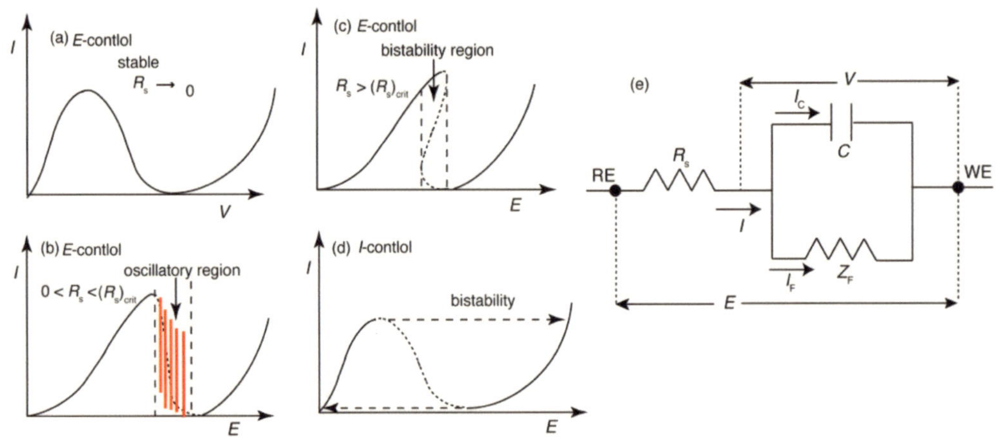

Fig. 1. Multisteady-state current-potential curves of N-type under potential controlled conditions with (a) a stable NDR-region at vanishing R_s, (b) current oscillations around NDR for intermediate values of R_s, (c) bistability without oscillations. (d) N-type curve under current-controlled conditions with bistability but without potential oscillations. (e) A general equivalent circuit of an electrochemical cell where *E* is the applied potential and *V* is the electrode potential.

Three basic categories are suggested to classify homogeneous (the spatial coupling is neglected) electrochemical oscillators. (Koper, 1996b; Krischer, 2003a):

1. N-NDR, characterized by an N-type current-potential curve for a vanishing ohmic resistance, $R_s \rightarrow 0$ (Fig. 1a). Potentiostatic current oscillations occur for intermediate

values of R_s (Fig. 1b) whereas bistability without oscillations exists for $R_s > (R_s)_{crit}$ (Fig. 1c). Galvanostatic potential oscillations do not occur (Fig. 1d). The majority of corroding metal | electrolyte systems that exhibit current oscillations across the active-to-passive transition are related to N-NDR systems. Among them the Fe | H_2SO_4, Fe | H_3PO_4 Co | H_3PO_4 and Zn | NaOH systems being few of them (Hudson & Tsotsis, 1994).

2. HN-NDR, characterized by a regime of a hidden negative differential resistance in the *I-E* polarization curve. Potentiostatic current oscillations around a regime of a positive slope occur when $R_s>(R_s)_{crit}$ whereas galvanostatic potential oscillations occur as well. Example of this category is the transpassive electrodissolution of Ni in H_2SO_4 (Lev et al., 1988).
3. S-NDR, characterized by an S-type polarization curve (Fig. 2). S-NDR systems oscillate under galvanostatic conditions at applied current values located within the NDR regime of the polarization curve. Example of this category may include the complicated dynamics of the electrodissolution of Fe in concentrated nitric acid (Gabrielli et al., 1976).

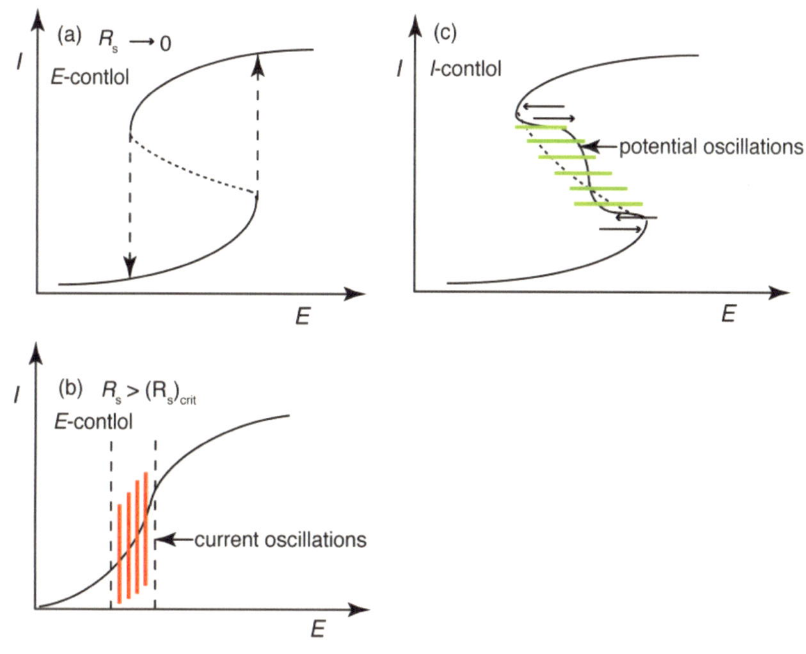

Fig. 2. Multisteady-state current-potential curves of S-type under potential-controlled conditions with (a) a vanishing R_s, (b) current oscillations for $R_s>(R_s)_{crit}$. (c) Potential oscillations under current-controlled conditions within the NDR regime, which in practice becomes not accessible and the *I-E* curve exhibits bistability.

As was shown in Fig. 1, the NDR in N-type current-potential curves is destabilized by increasing the ohmic resistance, R_s. Considering the general equivalent circuit (EC) of an electrochemical cell (Fig. 1e), the R_s represents the ohmic resistance, which includes un uncompensated cell resistance and a resistor connected in series between the working electrode, WE and ground. The total current *I* through the electrolyte interface consists of the faradaic current, I_F through the faradaic impedance, Z_F, and the current, I_C through the capacitor, *C* of the double layer. Under potential-controlled conditions, the potential, *E* between the WE and reference electrode (RE) should be constant and equal to $E=V + IR_s$. Destabilization of the N-type curves might be caused through the variation of the electrode potential, V. It introduces a positive feedback loop (activator) in the system by increasing R_s, which is used as a bifurcation parameter. Bifurcation parameter is a system parameter, which induces changes in the dynamic behavior of the system at a critical value. Dynamical changes occur through bifurcations (Koper, 1996b; Koper & Sluyters, 1993b).

Elucidation of the origin of oscillations includes the effect of the ohmic potential drop, IR_s and the discontinuous variation of the surface coverage ratio with the electrode potential due to the formation-dissolution of anodic surface films or the presence of autocatalytic chemical reactions concurrently occurring with electron transfer reactions. Mechanistically, the faradaic impedance, Z_F, depicted in the EC of an electrochemical cell (Fig. 1e), is related to the electrochemical processes at the metal (WE)|interface. Z_F should be derived from a reaction scheme. The basic equations are the mass balance for the reaction intermediates and charge balance equations derived from the general EC shown in Fig. 1e.

$$\frac{dc_i}{dt} = f_i(c_i, V) \tag{1}$$

$$\frac{dV}{dt} = \frac{I - I_F(V)}{CA} \tag{2}$$

where c_i is the surface concentration of reaction intermediates and A is the surface area. At least one intermediate species, which introduces a negative feedback loop (inhibitor), is required for an N-NDR system to exhibit periodic current oscillations. Details on this issue can be found in several comprehensive reviews and related articles (Koper, 1996b; Krischer, 1999; Krischer, 2003b).

As was mentioned above, on the basis of certain essential dynamical features, the Fe|H_2SO_4 system can be classified in the N-NDR category (Sazou et al., 1993a) where most of the metal|electrolyte systems belong. Therefore, only potentiostatic current oscillations are anticipated within a fixed potential region (Fig. 1b), when the IR_s exceeds an upper critical value, $IR_s > (IR_s)_{crit}$ bistability is expected, without oscillations. Under current-controlled conditions oscillations are not anticipated but only bistable behavior (Fig. 1d). The halide-perturbed Fe|H_2SO_4 system cannot be readily classified in one of the above categories and there is not doubt that it deserves a further investigation within this context. However, its dynamical behavior observed at relatively low chloride concentrations bears resemblance with the essential features of the HN-NDR oscillators (Krischer, 2003b). It seems, that oxide growth causes the NDR, whereas the slower action of chlorides on the oxide film and the gradual increase of passive-state current inhibits the appearance of NDR (Sazou et al., 2009).

4. Electrochemical behavior of the Fe|H_2SO_4 system

Fig. 3a illustrates the anodic current-potential (I-E) polarization curve of the Fe|0.75 M H_2SO_4 system traced under potential-controlled conditions at dE/dt=2 mV s^{-1} during both the forward and reverse backward potential scans. It seems that a variety of physico-electrochemical processes occur upon increasing/decreasing the potential within the region between -0.5 - 2.5 V.

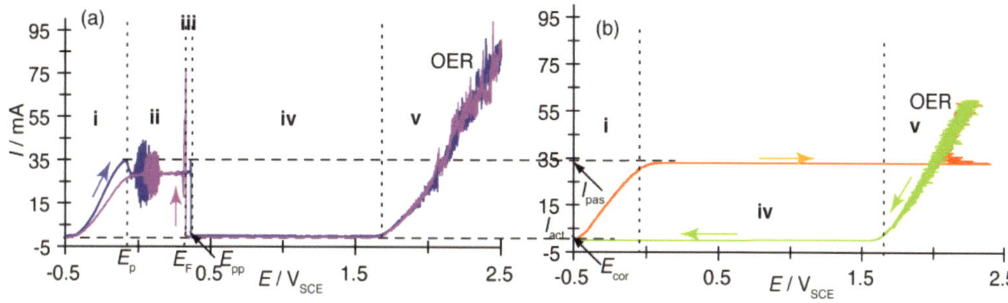

Fig. 3. (a) Potentiodynamic $I = f(E)$ curve traced at $dE/dt = 2$ mV s^{-1} and (b) galvanodynamic $E = f(I)$ curve traced at $dI/dt = 0.05$ mA s^{-1} of the Fe | 0.75 M H_2SO_4.

As can be seen in Fig. 3a, five typical regions are distinguished in the potentiodynamic *I-E* curve:

i. Active electrodissolution region, where Fe dissolution occurs from a film-free Fe surface through multiple stages (Keddam et al., 1984) and an overall reaction ,

$$Fe + nH_2O \rightarrow [Fe(H_2O)]^{2+} + 2e \qquad \text{(I)}$$

ii. Limiting current region (LCR) where, under proper conditions, current oscillations may occur (Geraldo et al., 1998; Kiss et al., 2006; Kleinke, 1995; Sazou et al., 2000b; Sazou et al., 2000c; Sazou & Pagitsas, 2006b) beyond the peak potential, E_p and before the establishment of a steady transport-controlled LCR within which formation-dissolution of the ferrous salt, $FeSO_4.7H_2O$ proceeds at equal rates, according to the overall reaction,

$$Fe^{2+} + SO_4^{2-} + 7H_2O \leftrightarrow FeSO_4 \cdot 7H_2O \qquad \text{(II)}$$

iii. Active-to-passive transition region, as defined during the forward potential scan (or passive-to-active transition region defined during the backward scan), associated with a hysteresis loop since transition to passivity, during the forward potential scan, occurs at the primary passivation potential, E_{pp} whereas reactivation of the Fe, during the backward scan, occurs at the Flade potential, E_F (Rush & Newman, 1995). The E_F and not the E_{pp} is considered as the passivation potential of the Fe electrode since the ohmic potential drop, IR_s becomes almost zero at E_F due to the very low current value established in the passive state traced during the backward scan. Therefore, the E_F determined in $I = f(E)$ curves is very close to the electrode potential V, which is related to the applied potential, E via the relationship, $E = V + IR_s$. In practice, R_s includes any series resistance added to the general EC of the electrochemical cell (Fig. 1e).

iv. The passive region, located between the E_F and transpassivation potential, E_{tr}. Transition of Fe to passivity can be represented by the overall reaction ,

$$Fe + x/2H_2O \rightarrow FeO_{x/2} + xH^+ + xe \qquad \text{(III)}$$

v. The transpassive region, extended beyond the E_{tr}, where the oxygen evolution reaction (OER) occurs (Sazou et al., 2009).

The galvanodynamic *I-E* curve of the Fe|0.75 M H_2SO_4 system (Fig. 3b) differs from the corresponding potentiodynamic *I-E* curve (Fig. 3a) in that region **ii** is not recorded. A sudden transition to passivity and, in turn, to OER (region **v**) occurs during the forward current scan, while the LCR (region **iii**) is skipped. It seems that the sudden active-to-passive transition occurs once the ferrous salt layer is established at the critical current value, I_{pas}. The I_{pas} corresponds to the peak potential E_p of the potentiodynamic curve (Fig. 3a). During the backward current scan, Fe sustains passivity (region **iv**) up to the corrosion potential, E_{cor} whereas the passive-to-active transition occurs at a very low current, I_{act}. A hysteresis loop exists because $I_{pas} \neq I_{act}$. Potential oscillations are never observed under galvanostatic conditions at any applied current value up to 60 mA (Sazou et al., 2011; Sazou et al., 2009), in line with the galvanodynamic curve (Fig. 3b).

On the contrary, periodic current oscillations occur under potentiostatic conditions, immediately after switching on the potential within the oscillatory region, ΔE_{osc} at $E < E_F$ (ΔE_{osc} = 30-35 mV for the Fe|0.75 M H_2SO_4 system). Typical potentiostatic current oscillations occurring within the ΔE_{osc} are illustrated in Figs. 4a-c.

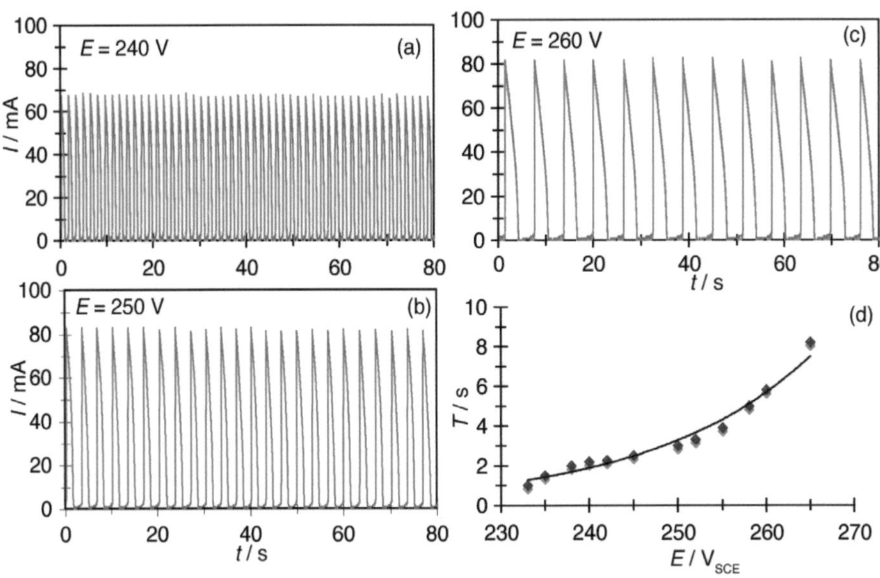

Fig. 4. (a-c) Potentiostatic current oscillations traced at different values of the applied potential, *E* and (d) dependence of the oscillation period, *T* on *E* for the Fe|0.75 M H_2SO_4 system.

Single periodic relaxation oscillations are revealed with a potential-dependent period, *T*. Fig. 4d shows that *T* increases upon increasing the potential (Podesta et al., 1979; Sazou et al., 1993a; Sazou & Pagitsas, 2003b). This indicates that the stability of the passive oxide film increases upon increasing the potential as a result of the increase of the oxide-film thickness and the concentration of Fe^{3+} in the oxide lattice (Engell, 1977; Vetter, 1971). The composition of the iron oxide film is related to Fe_3O_4 and γ-Fe_2O_3 (Toney et al., 1997) and its stability is determined roughly by the ratio c_{Fe3+} / c_{Fe2+}. Increasing the $c_{Fe^{3+}} / c_{Fe^{2+}}$ ratio within the oxide film results in an increase of the oxide stability in acid media (Engell, 1977). Thus upon increasing the potential at $E > E_F$, the oxide structure is related rather with the stable γ-Fe_2O_3 than with the less stable in acidic solutions Fe_3O_4 prevailing at $E < E_F$ where oscillations may occur.

The mechanism of spontaneous current oscillations of the Fe|0.75 M H_2SO_4 system is understood on the basis of the early suggested Franck-FitzHugh (F-F) model (Franck, 1978; Franck & Fitzhugh, 1961) according to which periodic passivation/activation of the Fe occurs due to local *pH* changes (Rush & Newman, 1995; Wang & Chen, 1998) that lead to a shift of the E_F with respect to the electrode potential, *V* because the E_F depends on the *pH*,

$$E_F = 0.58\text{-}0.058pH \text{ vs. NHE at 293 K} \quad (3)$$

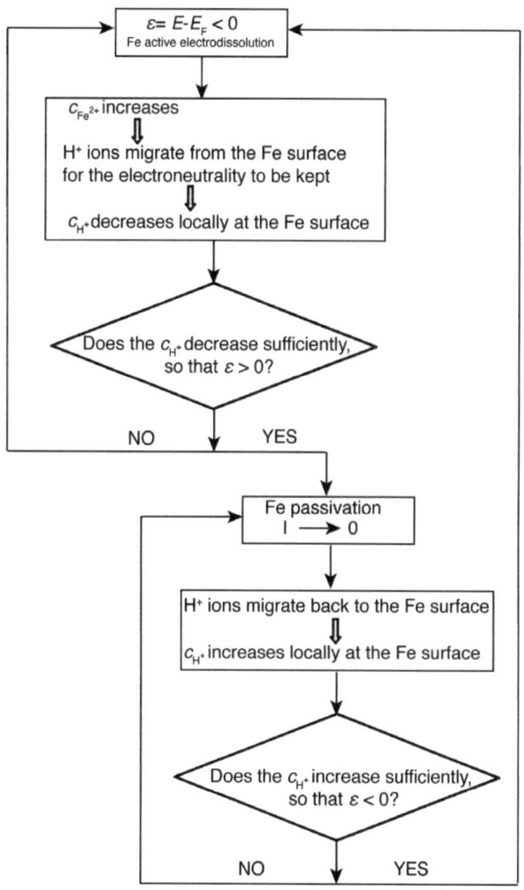

Fig. 5. Flow diagram illustrating the principle physico-electrochemcial processes involved in a current oscillatory cycle of the Fe|0.75 M H_2SO_4 system.

Furthermore, the formation of the ferrous salt layer and the ohmic potential drop, IR_s should be also taken into account for a realistic description of the periodic oscillations arisen across the passive-to-active transition of Fe in acid solutions (Birzu et al., 2001; Birzu et al., 2000; Koper & Sluyters, 1993a; Pagitsas & Sazou, 1991; Rush & Newman, 1995). As was mentioned above, the electrode potential, V coincides with the applied potential, E at E_F. Therefore, $E - E_F = \varepsilon$ denotes the difference from the passivation potential. It becomes evident that at $\varepsilon < 0$, the Fe active electrodissolution via the overall reaction (I) occurs. Inversely, at $\varepsilon > 0$, passivation of the Fe via the overall reaction (II) occurs. In between, the LCR is esrablished through the reaction (III) (Pagitsas et al., 2003). These processes, identified in the $I = f(E)$ of the Fe|0.75 M H_2SO_4 system (Fig. 3), are in practice the principle physico-electrochemical processes occurring during an oscillation cycle. Identical processes are also taken into account in improved versions of the F-F model (Koper & Sluyters, 1993a; Krischer, 2003a) as can be seen in the flow diagram displayed in Fig. 5.

5. Chemical perturbation of the Fe|H_2SO_4 system by adding small amounts of halides

In a series of studies, was shown that addition of halides, X ≡Cl^-, Br^-, I^- gives rise to changes in both the potentiodynamic I=f(E) (Pagitsas & Sazou, 1999; Sazou et al., 2000a; Sazou et al., 2000b; Sazou & Pagitsas, 2003b; Sazou et al., 1993b; Sazou et al., 1992; Sazou et al., 2009) and galvanodynamic E = f(I) curves (Sazou et al., 2011; Sazou et al., 2009) of the Fe|0.75 M H_2SO_4 system. All halide-induced changes can be identified approximately within regions **i-v** (Fig. 3). This indicates that halides participate in both the electrodissolution and passivation processes of Fe. However, since this article focuses on features that might be exploited to characterize pitting corrosion, special emphasis is placed on the passive and passive-active transition states of Fe.

The halide-induced changes together with nonilinear phenomena are investigated first on the basis of potentiodynamic, I = f(E) and galvanodynamic, E = f(I) curves. These curves can be considered as characteristic curves of the nonlinear system under study, in an analogy with the semiconductor "characteristic curve" used in solid state physics, or as a one-parameter "phase diagram" or "bifurcation diagram" in terms of nonlinear dynamics. Characteristic I = f(E) and E = f(I) curves exhibit all transitions between different steady and oscillatory states upon varying the applied potential acting as a bifurcation parameter. Then, the different states of the system being known, current or potential time-series are recorded under potentiostatic or galvanostatic conditions, respectively, at potentials or current values located within the corresponding oscillatory regions. A slight deviation is noticed in determining the upper and lower limits of the oscillatory region under static conditions as compared with dynamic I = f(E) and E = f(I) curves.

5.1 Under potential-controlled conditions

The effect of an increasing chloride concentration, within a relatively low-concentration range ($c_{Cl^-} < 20$ mM), on potentiodynamic I= f(E) curves is displayed in Fig. 6.

Inspection of the I= f(E) curves of Fig. 6, reveal that pitting corrosion manifests itself in changes that are summarized as follows:

1. The halide-induced oscillatory region, $\Delta E_{osc,Cl}$, relative to the halide-free ΔE_{osc}, is extended towards higher potentials ($\Delta E_{osc,Cl} > \Delta E_{osc}$).

2. The lower, E_{low} and upper, E_{upp} potential limits of the oscillatory region shift towards higher values indicating destabilization of the stable passive state existing in the halide-free system. It is found that both E_{low} and E_{upp} vary linearly with the log(c_{Cl^-}). Appropriate analysis, leads to the critical values of c_{Cl^-}, beyond which oxide formation becomes unlikely and hence transition to a mass-transport LCR may occur due to the formation-dissolution of ferrous salts signifying a stable pit growth. This value is found to be ~30 mM in agreement with experimental observations (Sazou et al., 2000a; Sazou et al., 2000b).
3. The current in the passive state increases. It is lower during the forward potential scan ($I_{pas,f}$ in Fig. 6b) than during the backward one ($I_{pas,b}$ in Fig. 6c). This can be interpreted considering that pitting corrosion is a dynamical process and therefore, the longer time elapsed for the backward scan allows the progress of pit propagation and/or growth to a greater degree than during the forward scan.
4. The maximum oscillatory current, $(I_{osc})_{max}$ (Fig. 6b) deviates from the kinetics of the linear segment in region **i**, indicating a larger real surface of the Fe electrode. This is attributed to the increase of the surface roughness due to pitting corrosion as compared with the uniformly corroding Fe surface in the halide-free system (Fig. 6a). The magnitude of the deviation is expressed as the ratio $(I_{osc})_{max}/(I_{osc})_{max,\ exp}$, where the $(I_{osc})_{max,\ exp}$ is the current expected on the basis of the relationship $E = V + IR_s$. The latter relationship is valid in the linear segment of the active region **i** located beyond the Tafel region (Pagitsas et al., 2007; Pagitsas et al., 2008).
5. No access to E_{tr} is possible whereas the critical pitting potential E_{pit} appears (Fig. 6b). The E_{pit} is the critical potential for stable pitting to occur.

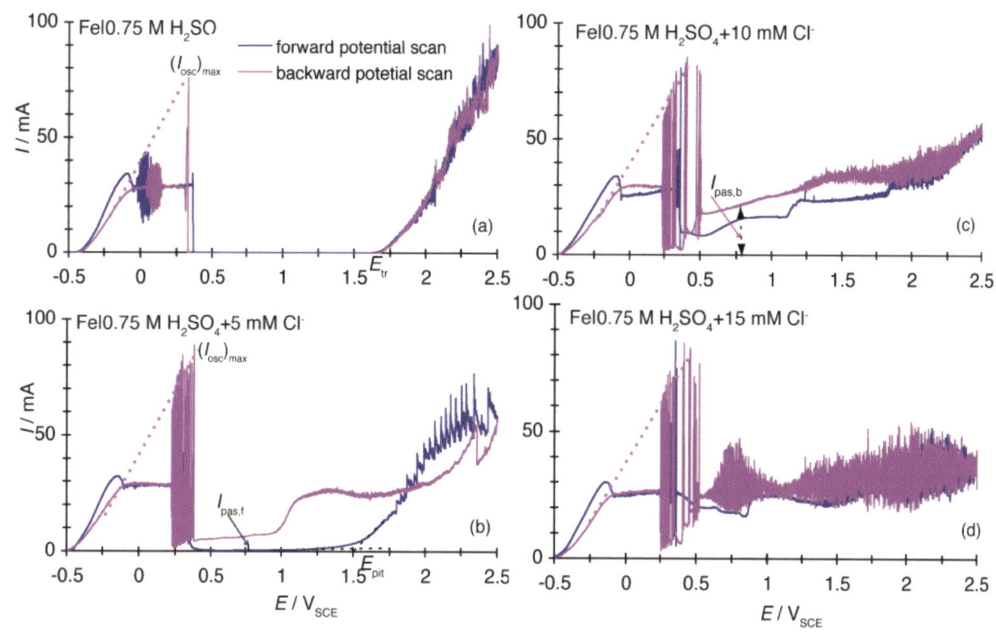

Fig. 6. Chloride-induced changes in the potentiodynamic $I = f(E)$ curves of the Fe | 0.75 M H_2SO_4 system traced at dE/dt = 2 mV s^{-1}.

Moreover, Fig. 6 shows that increasing gradually the c_{Cl^-} the current in the passive state increases (Table 1). At $c_{Cl^-} > 15$ mM, both $I_{pas,f}$ and $I_{pas,b}$ tend to reach a limiting value within the potential region between ~0.3 and 2.5 V whereas new oscillatory states emerge out of the passive state. These current oscillations are associated with the precipitation-dissolution of ferrous salt layers in front of pits grown on the Fe surface, while the OER rate diminishes (Sazou et al., 2000b; Sazou & Pagitsas, 2003a).

Adding Br^- and I^- ions leads to similar changes in the corresponding potentiodynamic *I-E* curves with those mentioned above. However, comparing quantities such as the $(I_{osc})_{max}$ / $(I_{osc})_{max,\ exp}$, ΔE_{osc} and the current in the passive state allows characterization of the extent of pitting corrosion induced by each halide ion (Sazou et al., 2000a). This becomes obvious from Table 1, which summarizes the values of these quantities for all halides in comparison with those obtained for the halide-free Fe|0.75 M H_2SO_4 system. At relatively low halide concentrations, these quantities depend on c_{X^-} and the aggressiveness of halides. It is thus concerned with localized oxide breakdown and repeated activation-repassivation events of the entire Fe-disc surface at early stages of pitting corrosion. However, depending on the halide identity, additional individual differences are noticed in the case of I^- and fluoride species. In the former case, it is observed that $I_{pas,f} > I_{pas,b}$, which is an inverse relationship compared to that anticipated for pitting corrosion induced by Cl^- and Br^-. This is assigned to the formation of a compact iodine layer over the Fe surface due to iodide electrochemical oxidation. Iodine layer seems to prevent the evolution of pit growth. In he case of fluoride species, though the ΔE_{osc} is extended towards higher potentials, drastic changes in the $I_{osc,\ max}$ and the current in the passive state are not observed indicating an enhanced general corrosion instead of pitting.

Addition	c (mM)	ΔE_{osc} (mV)	$I_{pas,\ f}$(mA)	$I_{pas,\ b}$(mA)	$(I_{osc})_{max}/(I_{osc})_{max,\ exp}$	E_{tr} , E_{pit} (V)
None	-	235-270	0.15	0.15	1.01	1.65
NaF	10	240-290	0.22	0.22	1.05	1.65
	20	245-310	0.25	0.25	1.03	1.65
NaCl	10	240-380	2.4	13.7	1.2	1.00
	20	290-520	23.7	22.4	1.23	LCR
NaBr	10	255-500	3.9	3.7	1.13	1.4
	20	280-700	22.5	22.5	1.18	LCR
NaI	10	243-440	6.8	1.12	1.01	1.55
	20	245-450	5.5	0.6	1.02	1.37

Table 1. Effect of halide ions, X^- on the oscillatory potential region ΔE_{osc}, the current in the passive state observed during the forward, $I_{pas,f}$ and backward, $I_{pas,b}$ potential scans, the maximum oscillatory current ratio $(I_{osc})_{max}/(I_{osc})_{max,\ exp}$ and the transpassivation potential, E_{tr} or the pitting potential, E_{pit} appeared in the presence of X^-.

Besides changes observed in potentiodynamic I = f(E) curves, pitting corrosion manifests itself in potentiostatic current oscillations too. An example of halide-induced oscillations is given in Fig. 7a. Fig. 7a displays a transition between single periodic oscillations observed in the halide-free system, immediately after switching on the potential at $E < E_F$, and the halide-induced complex periodic oscillations appeared after an induction period of time, t_{ind} (Fig. 7b). The halide-induced current oscillations occur over a wide potential region (Table 1) and their periodicity was found to follow period doubling, quadrupling and aperiodicity by increasing the applied potential, E and c_{X^-} (Sazou et al., 2000b). Depending on E and c_{X^-} different temporal patterns were recorded such as bursting and beating. Variation of the current waveform was also observed at longer times as it is anticipated for pitting corrosion, being strongly- dependent on time (Sazou et al., 2000b; Sazou et al., 1993b).

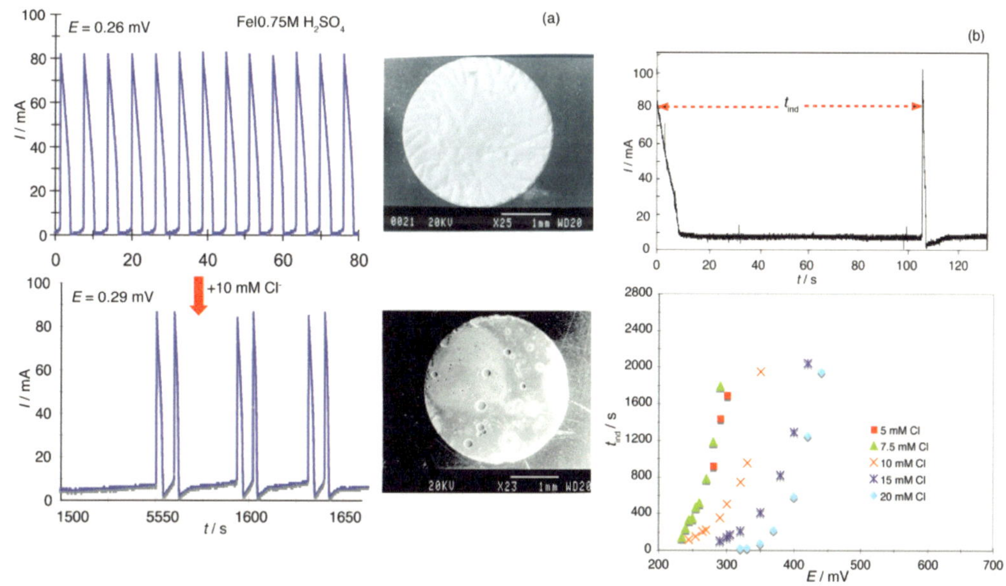

Fig. 7. (a) Transition of the single periodic to a complex periodic current oscillation induced by adding Cl^- into the Fe | 0.75 M H_2SO_4 system. Next to each oscillation waveform, SEM micrographs display the corresponding Fe surface morphologies. (b) Induction time, t_{ind} occurring prior the onset of chloride-induced oscillations and dependence of t_{ind} on the potential at various c_{Cl^-}

The complex oscillations induced by Cl^- as well as Br^- and I^- is the result of the aggressive action of halides on the Fe surface (Sazou et al., 2000a; Sazou et al., 2000b). This is confirmed by SEM observations, an example of which is depicted in Fig. 7a. The morphology of the Fe-disc surface during the occurrence of single periodic oscillations reflects a general corrosion induced by H^+ ions. Hydrogen ions catalyze the uniform dissolution of the passive oxide film consisting primarily of Fe_3O_4 (Engell, 1977; Sato, 1990). Associated with temporal current patterning, are also spatial phenomena that deserve to be investigated (Hudson et al., 1993). On the other hand, when complex periodic current oscillations occur in the presence of chlorides and other X^- ions, the morphology of the Fe surface reveals hemispherical pits as a result of the local breakdown of passivity on Fe. Local active areas generated by the local action of halides results in an inhomogeneous passive Fe surface and perhaps in new spatiotemporal patterns.

It is noted that besides the potential, the solution resistance, R_s or equally a variable external series-resistance, R_{ex} inserted between the working and reference electrodes, the rotation speed, ω of the rotating Fe-disc electrode, the solution *pH* and temperature, all parameters control the nonlinear behavior of the Fe|0.75 M H_2SO_4 electrochemical oscillator. However, though these contol parameters influence the onset, the period and amplitude of oscillations, none of them changes the oscillation waveform. Whenever current oscillations appear across the passive-to-active transition region (at $E < E_F$), they are single periodic of relaxation type. To our knowledge, the type of these oscillations will change only when a halide-induced chemical perturbation of the passive state of the Fe| H_2SO_4 system is conducted and pitting instead of uniform corrosion occurs. This is a striking indication that new physico-electrochemical processes have been triggered by halides manifested in the variety of complex oscillations.

Inspection of Fig. 7b shows that a fluctuating steady current exist during the induction period of time, t_{ind} elapsed before the onset of oscillations. This is indicative of the aggressive action of Cl^-, which leads to the nucleation of small pits that in turn are repassivated immediately. It proceeds until a complete destabilization of the passive state occurs (1st activation event). The transition to the active state is followed by repeated passivation-activation events (complex oscillations) that constitute a phenomenon termed as unstable pitting corrosion. Therefore, the t_{ind} characterizes the kinetics of the oxide attack by X^- ions. It was found that t_{ind} depends on both the c_{X^-} and E. An example is given in Fig. 7b for a chloride- perturbed Fe|0.75 M H_2SO_4 system. As Fig. 7b shows, increasing the c_{X^-} leads to the decrease of t_{ind} indicating promotion of the local breakdown of the oxide film. On the contrary, increasing the applied potential the t_{ind} increases. The rate of unstable pitting corrosion diminishes by increasing the potential due to the enhancement of oxide stability.

Another quantity describing quantitatively the competitive process between halide action and enhancement of oxide stability is the oscillation period, *T*. As can be seen in Fig. 8a, *T* decreases with c_{Cl^-}, while it increases with *E*. It is noticed that the increase of *T* observed at lower c_{Cl^-} is interpreted in terms of changes in periodicity since period doubling and quadrupling occurs. More accurate empirical relationships are obtained if an average activation rate (number of spikes over a period of time) is employed instead of *T* (Pagitsas et al., 2001; Pagitsas et al., 2002; Sazou et al., 2000a).

In this context, it should be noted that the decrease of *T* with c_{Cl^-} is also associated with the general corrosion occurring concurrently with unstable pitting corrosion. As can be seen in Fig. 8b, a similar dependence of *T* on the concentration of fluoride species is also obtained. It is well known that fluorides in acid solutions (*pH* ~ 0-0.5) cause general corrosion but not pitting since HF is the largely predominant species, while other fluoride species may coexist at negligible amounts (Pagitsas et al., 2001; Pagitsas et al., 2002; Strehblow, 1995). In agreement with the fluoride effect, is also the effect of c_{H^+} on *T* (Sazou & Pagitsas, 2003b). As was mentioned above, the onset of current oscillations in the halide-free Fe|0.75 M H_2SO_4 system is assigned to destabilization of the oxide film due to *pH* changes occurring uniformly at the Fe surface (Eq. (3)). In this case, general corrosion of the oxide film is induced by H^+ ions only around the E_F (Fig. 3) and hence T decreases with increasing the c_{H+} (Sazou & Pagitsas, 2003).

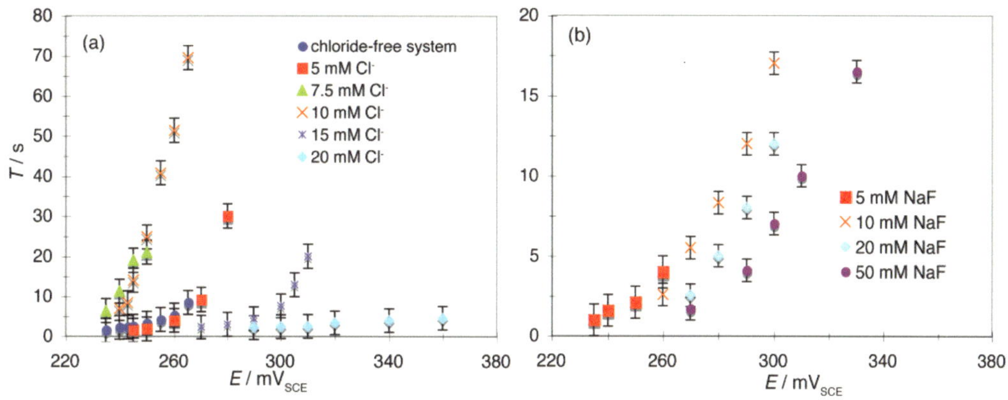

Fig. 8. Dependence of the oscillation period, T as a function of the applied potential, E at (a) various c_{Cl^-} and (b) various concentrations of fluoride species.

Therefore, distinction between pitting and general corrosion is possible on the basis of quantities obtained from both potentiodynamic I = f(E) and potentiostatic I = f(t) curves. Besides, Cl^-, Br^-, I^- and fluoride species, chlorates and perchlorates were also used in perturbing the Fe | 0.75 M H_2SO_4 system (Pagitsas et al., 2007; Pagitsas et al., 2008). The effect of chlorates and perchlorates on the Fe passive surface, being disputable in the literature, was clarified using all the above-mentioned diagnostic criteria including ΔE_{osc}, $I_{pas,f}$ and $I_{pas,b}$, $(I_{osc})_{max}/(I_{osc})_{max,\,exp}$, E_{pit}, t_{ind} and T.

Moreover, these diagnostic criteria were also tested in the presence of nitrates in chloride-containing sulfuric acid solutions (Sazou & Pagitsas, 2002). Newman and Ajjawi characterized the effect of nitrates on stainless steel as peculiar (Newmann & Ajjawi, 1986). Regarding pitting corrosion, nitrates may act either as activating or inhibiting species (Fujioka et al., 1996). This property of nitrates was also realized in the case of Fe. It is manifested in several features of current oscillations observed over a wide potential region. Appropriate analysis of I = f(E) and I = f(t) curves demonstrated that nitrates may stimulate pitting corrosion at lower potentials, while may cause a sudden passivation of Fe at higher potentials. This behavior is interpreted by taking into account the electrochemical and homogeneous chemical reactions of nitrates. Current oscillation seems to be a suitable probe to indicate both qualitatively and quantitatively if a stable passive state is established in corrosive media (Sazou & Pagitsas, 2002).

5.2 Under current-controlled conditions

Fig. 9 shows galvanodynamic E = f(I) curves of the Fe | 0.75 M H_2SO_4 system traced at gradually increasing c_{Cl^-}. Chlorides seem to induce:

1. Potential oscillations at a critical c_{Cl^-}.
2. Occurrence of more potential oscillations during the backward current scan as well as by increasing c_{Cl^-}.

3. Considerable increase of I_{act} with a slight decrease of I_{pas} by increasing c_{Cl^-}. Hence the width of the hysteresis loop, $\Delta I = |I_{pas} - I_{act}|$ decreases (Fig. 9a). The I_{pas} and I_{act} are defined as the critical current values where transition to passive and active states occurs during the forward and inversely backward current scans, respectively.

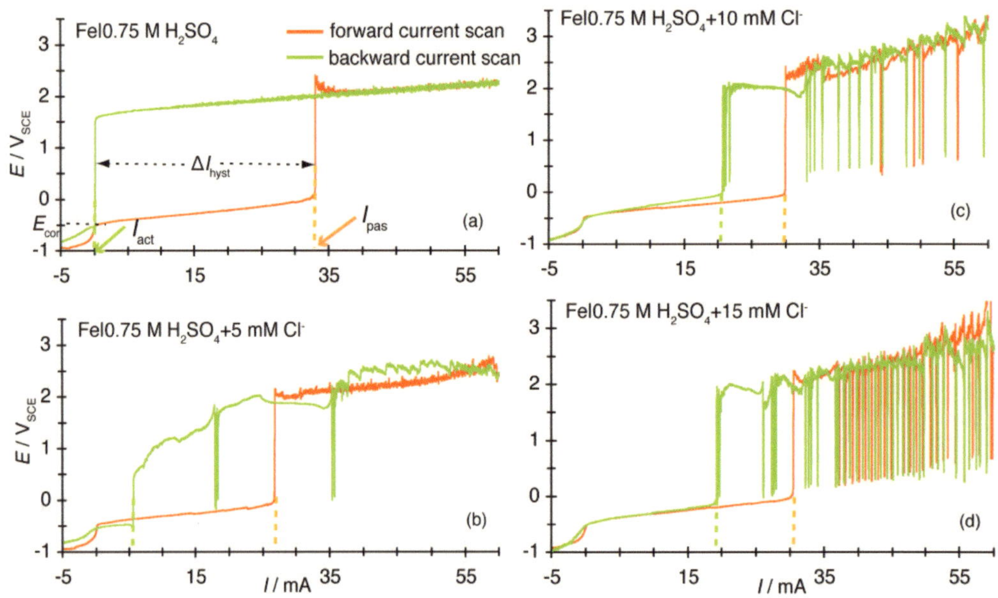

Fig. 9. Chloride-induced changes in the galvanodynamic $E = f(I)$ curves of the Fe|0.75 M H_2SO_4 system traced at $dI/dt = 0.05$ mA s^{-1}.

Apparently, potential oscillations in galvanodynamic $E = f(I)$ curves constitute manifestation of pitting corrosion since, as was mentioned in sections 3 & 4, no oscillations should occur for the halide-free Fe|0.75 M H_2SO_4 system under a current control. Corresponding changes in $E = f(I)$ curves are also induced by adding other halide species. An example of $E = f(I)$ curves at 20 mM of fluorides, Cl^-, Br^- and I^- ions is illustrated in Fig. 10.

Comparing the $E = f(I)$ curves illustrated in Fig. 10, it seems that fluorides do not induce potential oscillations, as it should be anticipated, since fluorides cause only general corrosion of the Fe surface. In the present case, general corrosion proceeds concurrently with the OER exemplifying itself by the occurrence of small amplitude potential fluctuations due to the formation and subsequent escape of oxygen bubbles from the Fe surface (Fig. 10a). Regarding Br^-, potential oscillation does appear (Fig. 10c), though to a lesser degree than in the presence of Cl^- (Fig. 10b), in agreement with the lesser aggressiveness of Br^- as compared with that of Cl^-.

Inspecting Fig. 10d, one may see that no pitting corrosion occurs in the presence of I^-. However, it is well known that iodide does cause pitting corrosion (Strehblow, 1995), manifested also in both the I=f(E) and I=f(t) curves (Pagitsas et al., 2002; Sazou et al., 2000a). This apparent discrepancy can be interpreted by taking into account oxidation processes of iodides that result in the formation of a solid iodine film on the Fe surface (Ma & Vitt, 1999;

Vitt & Johnson, 1992). This iodine film hinders the activation of the Fe surface expected to occur due to the localized action of I^-. Thus any noticeable increase of I_{act} is not shown. In fact, the I_{act} remains equal with that of the unperturbed system (Fig. 9a). Instead of large amplitude oscillations, indicative of localized corrosion, a new type of potential oscillation emerges (Fig. 10d) associated with the OER occurring concurrently with the formation-dissolution of the iodine layer. The features of these low amplitude oscillations are influenced by the c_{I^-} and applied current values and become more pronounced at higher c_{I^-} (~50 mM). Therefore, the electrochemical and chemical behavior of I^- together with their action on the passive Fe surface becomes very complicated under current-controlled conditions. Further investigation within a different context deserves to be carried out.

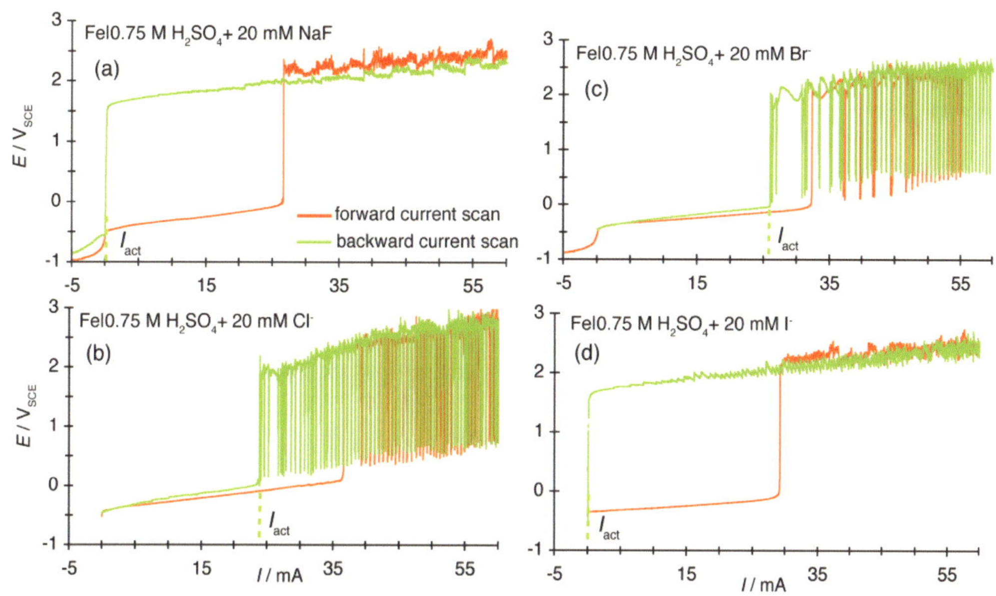

Fig. 10. Comparison of the effect of various halide species on the galvanodynamic $E = f(I)$ curves of the Fe | 0.75 M H_2SO_4 system traced at $dI/dt = 0.05$ mA s^{-1}.

Table 2 summarizes the values of I_{pas} and I_{act} obtained for the halide-free Fe | 0.75 M H_2SO_4 system in comparison with the corresponding values evaluated for the halide-perturbed one. The occurrence of potential oscillations and the quantity I_{act} are associated with pitting corrosion. The I_{act} increases by increasing either the halide concentration or the aggressiveness of halides implying stimulation of pitting corrosion. The higher the I_{act} or the lower the width of the hysteresis loop is, ΔI, the greater is the susceptibility of Fe to pitting corrosion. Comparing the aggressiveness of Cl^- and Br^- in terms of the I_{act} or ΔI, the order Cl^- > Br^- is found, in agreement with the order found from the nonlinear dynamical response obtained under potential-controlled conditions (Pagitsas et al., 2002; Sazou et al., 2000a), as well as with literature data based on other criteria (Janik-Czachor, 1981; Macdonald, 1992; Strehblow, 1995).

The current region within which potential oscillations are expected to occur at a constant applied current, I_{appl} can be deduced from the E=f(I) curves. As was mentioned in the beginning of this section, E=f(I) curves represent roughly one-parameter bifurcation diagrams. It seems that, for oscillations to occur, the I_{appl} should be approximately higher than I_{act}. Fig. 11 shows examples of galvanostatic E=f(t) curves traced for 20 min at I_{appl} = 30 mA for the halide-free and chloride-perturbed Fe | 0.75 M H_2SO_4 system at different c_{Cl^-}.

Addition	c (mM)	I_{pas} (mA)	I_{act} (mA)	$\Delta I= I_{pas}- I_{act}$
None	-	33	0.15	32.85
NaF	10	33	0.22	32.78
	20	32	0.25	31.75
NaCl	10	30	20	10
	20	29	25	4
NaBr	10	32	18	14
	20	29	22.6	6.4
NaI	10	24	0.15	23.85
	20	28.9	0.15	28.75

Table 2. Effect of halides, X^- on the current, I_{pas} at which transition to passivity occurs during the forward current scan, the current I_{act}, where reactivation occurs during the backward current scan and the width of the hysteresis loop, ΔI defined from galvanodynamic curves (dI/dt=0.05 mA s^{-1}) of the Fe | 0.75 M H_2SO_4 system.

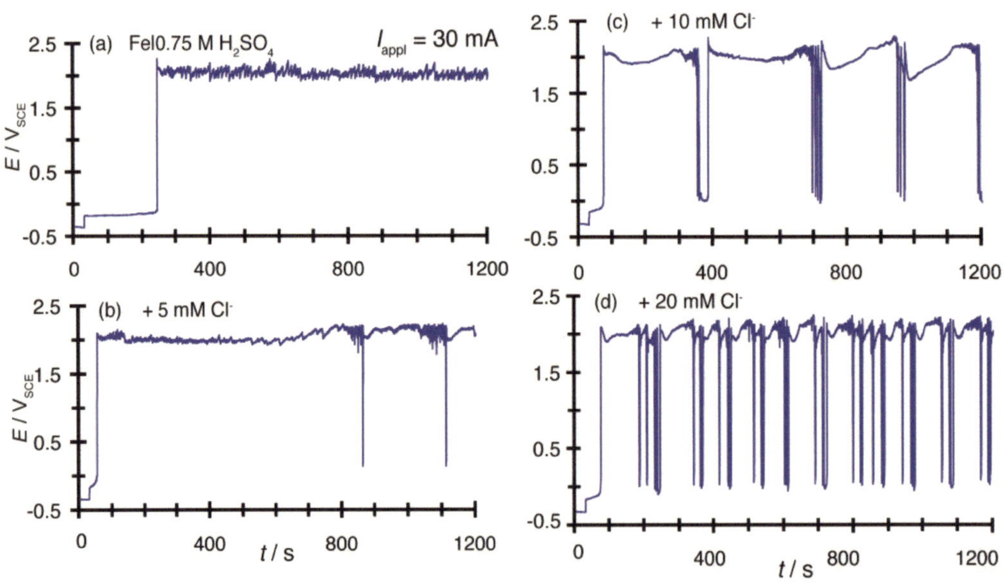

Fig. 11. Chloride-induced potential oscillations of the Fe | 0.75 M H_2SO_4 system traced under galvanostatic conditions at I_{appl}=30 mA.

Similar potential oscillations, with those illustrated in Fig. 11, were also observed in the presence of Br^-. In summary, chloride- and bromide-induced changes in galvanostatic $E=f(t)$ curves at constant I_{appl} for various c_{Cl^-} and at constant c_{Cl^-} for various I_{appl} include:

1. Onset of potential oscillations of large amplitude (~2 V) when c_{Cl^-} is higher than a critical value (>5 mM) and only if $I_{appl} > I_{act}$.
2. The potential oscillates between the two steady states, namely the passive and active states. This indicates that initiation of pitting results in the destabilization of passivity on Fe and activation of the entire Fe surface.
3. Different waveforms of potential oscillations depending on c_{Cl^-} (or c_{Br^-}) and I_{appl} with an average frequency that increases with increasing c_{Cl^-} and I_{appl}.
4. Occurrence of certain induction period of time, t_{ind} before oscillations start, which decreases by increasing the c_{Cl^-} and I_{appl}.

The dependence of t_{ind} and average oscillation frequency with c_{Cl^-} and I_{appl} is displayed in Figs. 12a, b. The oscillation frequency is expressed as the average firing rate, $\langle r \rangle$ defined by the ratio, $\langle r \rangle = N/\Delta\tau$, where N is the number of spikes (passive-active events) appeared during a fixed duration, $\Delta\tau$ of the experiment (Dayan & Abbott, 2001). The N is measured at $t > t_{ind}$ (Sazou et al., 2009). It becomes clear that t_{ind} reflects the kinetics of pit initiation on the passive Fe surface whereas $\langle r \rangle$ is rather related to the pit growth and propagation. Therefore, both t_{ind} and $\langle r \rangle$ can be used to describe quantitatively pitting corrosion on passive Fe. The quantities t_{ind} and $\langle r \rangle$ are currently used to estimate the inhibiting effect of nitrates on pitting corrosion.

When Fe is in the passive state (high-potential state) and chlorides start their action generating local active areas on the Fe surface, the oxide becomes gradually dark brown due to the conversion of its outer layer into ferrous oxo-chloride complexes. At the moment of Fe activation, all anodic layers, being separated from the Fe substrate, seem streaming away from the electrode. Due to the high current, the active Fe surface abruptly passivates and correspondingly the potential increases to its highest value. SEM images reveal an inhomogeneous growth of the passive oxide since it covers both localized activated and perhaps never-activated sites.

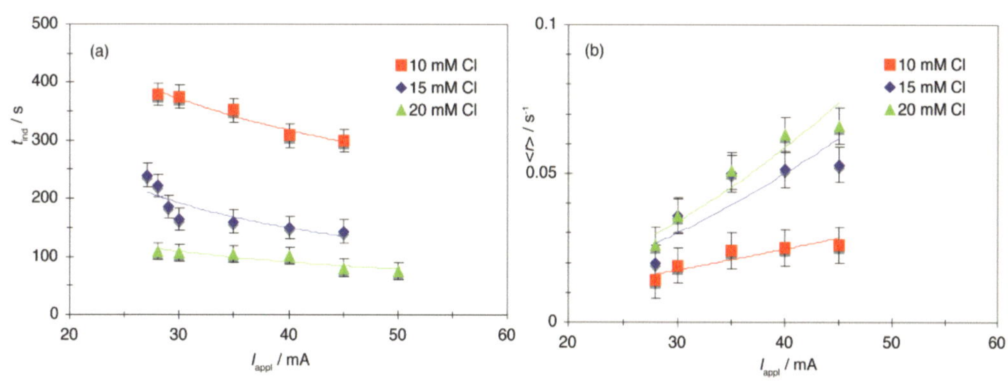

Fig. 12. Dependence of the (a) induction time, t_{ind} required for potential oscillations to start and (b) average firing rate, $\langle r \rangle$ as a function of the applied current, I_{appl} and c_{Cl^-} for the chloride-perturbed Fe | 0.75 M H_2SO_4 system.

The mechanism of passive-active oscillations associated with unstable pitting corrosion includes the formation and detachment of the oxide film that can be sufficiently explained in terms of the point defect model (PDM) (Macdonald, 1992; Pagitsas et al., 2001; Pagitsas et al., 2002; Pagitsas et al., 2003; Sazou et al., 2009). PDM is a realistic quantitative model that includes many of the oxide properties and explains many of the experimental observations during oxide growth and its breakdown. The processes leading to pitting corrosion are associated with the occupation of oxygen vacancies by halides, X^-. This reaction results in perturbation of a Schottky-pair equilibrium and autocatalytic generation of cation vacancies. Cation vacancies accumulate at the Fe | oxide interface leading to the formation of void and separation of the oxide from the Fe substrate. Simultaneously, the thickness of the oxide film decreases due to general corrosion through the formation of surface complexes between iron lattice-cations and halides. When the void exceeds a critical size and the oxide film over the void thins below a critical thickness film breakdown occurs at this particular site (Macdonald, 1992; Sazou et al., 2009).

At sufficiently high c_{Cl^-} or c_{Br^-}, dissolution rates are enhanced whereas oxide formation becomes unlikely. Instead, formation of ferrous salt layers is facilitated leading to the electropolishing dissolution state (Li et al., 1993; Li et al., 1990). This situation is discussed briefly below only in the case of potential-controlled conditions.

6. Non-linear dynamical response of the Fe|H_2SO_4 system at relatively high concentrations of halides

Fig. 13a shows that at relatively high halide concentrations (i.e. $c_{Cl^-} > 20$ mM), Fe cannot sustain passivity and a limiting current region (LCR) is established out of the passive state within 0.3 and 2.7 V (the upper potential limit used in the potentiodynamic measurements). This LCR is due to the precipitation-dissolution of salt layers since the oxide growth is prevented by Cl^- and differs from the LCR appeared for $E < E_F$, where oxide formation is thermodynamically prohibited. Within this "new" LCR, two distinct types of current oscillations are observed.

- Type I, called also as passive-active oscillations appeared within the lower potential regime ($E < 0.6$ V) either as a continuous spiking (beating) or aperiodic bursting. These oscillations arise out of a limiting current state with a full-developed amplitude and differ from those observed at relatively low c_{Cl^-} , which arise out of a passive state (Fig. 7).
- Type II, chaotic oscillations of a relatively small amplitude occurring at higher potentials ($E > 0.6$ V). The extent of each oscillatory regime depends on the halide concentration and halide identity. Upon a further increase of c_{Cl^-}, the regime corresponding to oscillations of type I is restricted gradually. For $c_{Cl^-} > 40$ mM, current oscillations of type II dominate the entire LCR for $E > 0.3$ V (Fig. 13a).

An induction period of time, t_{ind} is elapsed before current oscillations of type I or II appear. During t_{ind}, the current reaches a steady state value during which precipitation-dissolution of ferrous salts occurs at equal rates. Precipitation of ferrous salt occurs inside pits when a local supersaturation condition for Fe^{2+} and sulfates/chlorides is reached. There are evidences (Sazou & Pagitsas, 2003a) that the bifurcation potential, E_{bif}, for the transition from oscillations of type II to those of type I coincides with the repassivation potential, E_R

used in pitting corrosion studies under steady state conditions (Sato, 1987; Sato, 1989). E_R it is the critical potential at which a transition from a polishing state dissolution (bright pits) to active state dissolution (etching pits) occurs. Critical conditions for the onset of different types of oscillations may be defined in terms of the critical pit solution composition (critical c_{Cl^-} and c_{H^+}) at which Fe cannot sustain passivity and, thereby, pit stabilization is possible (Sazou & Pagitsas, 2003a).

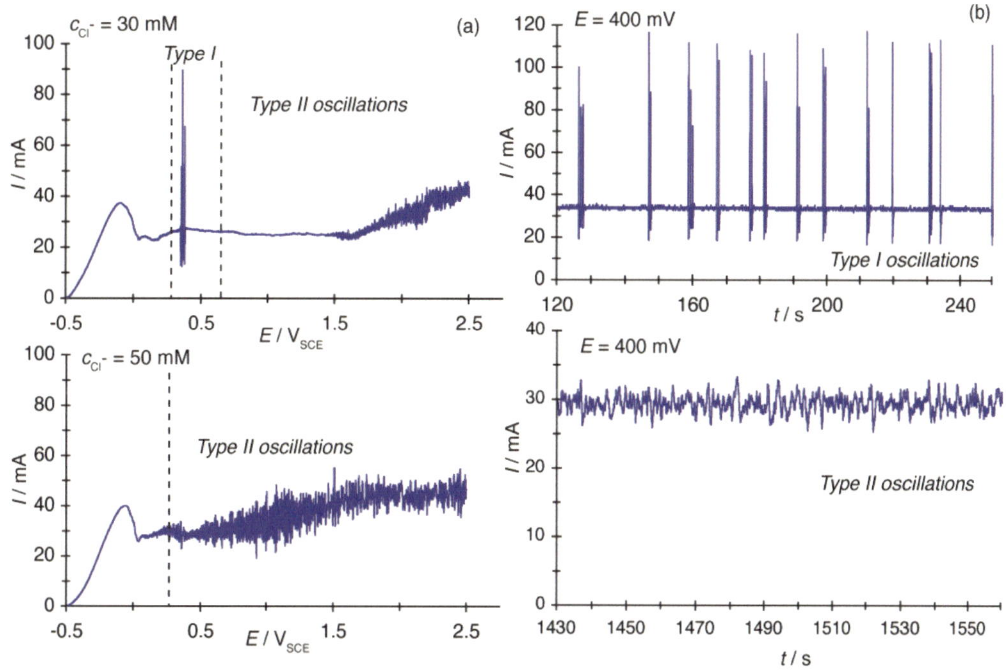

Fig. 13. (a) Non-linear dynamical response of the Fe | 0.75 M H_2SO_4 system in the presence of relatively high c_{Cl^-}. It exhibits a transition from a situation where oscillations of type I and II appear in the *I-E* curves to one where oscillations of type II dominate the whole LCR at $E >$ 0.3 V. Polarization curves were traced at $dE/dt = 2$ mV s^{-1}. (b) Representative examples of type I and type II current oscillations corresponding to the I=f(E) curves shown in (a).

An example of a potential-induced transition between potentiostatic current oscillations of type I (aperiodic bursting) to those of type II (low amplitude chaotic oscillations) is illustrated in Fig. 14 for the Fe | 0.75 M H_2SO_4 + 30 mM Cl^- system. It seems that this transition occurs around $E_{bif} = 0.55$ V which coincides with the E_R (Sazou & Pagitsas, 2003a).

Oscillations of type II occurring at either high potentials of the oscillatory region at relatively low c_{Cl^-} or within the entire oscillatory region at sufficiently high c_{Cl^-} originate from processes similar to those responsible for chaotic oscillations observed at the beginning of the LCR ($E < 0.3$ V) shown in the I = f(E) curve of the halide-free Fe | 0.75 M H_2SO_4 system (Fig. 3a) (Sazou & Pagitsas, 2006b). Supersaturation conditions of the ferrous salts established inside pits results in a density gradient Δd between the solution in the interfacial regime in front of the Fe electrode and the bulk solution. When Δd exceeds a critical value the steady limiting current becomes unstable. This condition is fulfilled in the presence of a critical *IR* drop (Georgolios & Sazou, 1998; Pickering, 1989; Pickering & Frankenthal, 1972).

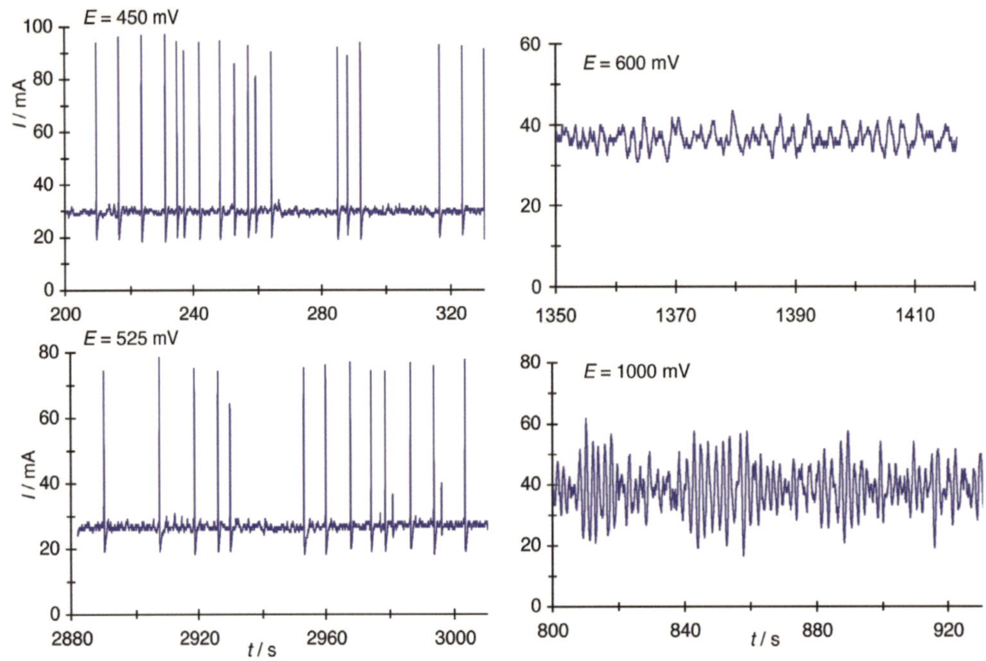

Fig. 14. Sequence of current oscillations at late stages of pitting corrosion of Fe in 0.75 M H_2SO_4 + 30 mM Cl^- displaying a transition from oscillations of type I to those of type II upon increasing the applied potential, *E*.

7. Alternate diagnostic criteria to characterize pitting corrosion at early stages of pitting corrosion

It becomes clear that the non-linear dynamical response of the halide-perturbed Fe|0.75 M H_2SO_4 system exemplified either under potential- or current-controlled conditions reflects the aggressive action of halide ions, especially of Cl^- on the Fe passive oxide film. Steady-state processes leading to passive and active states of Fe in a halide-free sulfuric acid solution are perturbed through a series of physico-electrochemical reactions including autocatalytic steps. In fact, pit nucleation, propagation and growth are autocatalytic processes (Budiansky et al., 2004; Lunt et al., 2002; Macdonald, 1992). Pit repassivation or stable growth can be realized by investigating the system oscillatory states and oscillation waveform. Therefore, oscillations might be used like a "spectroscopic" technique to detect pitting corrosion and moreover to characterize unstable and stable stages during pit evolution. A summary of oscillatory phenomena expected to arise at different stages of pitting corrosion can be seen in the flow diagram displayed in Fig. 15.

Under potential-controlled conditions, the nonlinear dynamical response of the halide-perturbed Fe|0.75 M H_2SO_4 system recorded in I=f(E) and I=f(t) curves is characterized by complex current oscillations. The halide concentration, c_{X^-}, applied potential, E and time, all affect characteristic features of oscillations, which point to pit initiation, propagation, and growth on an otherwise passive Fe surface.

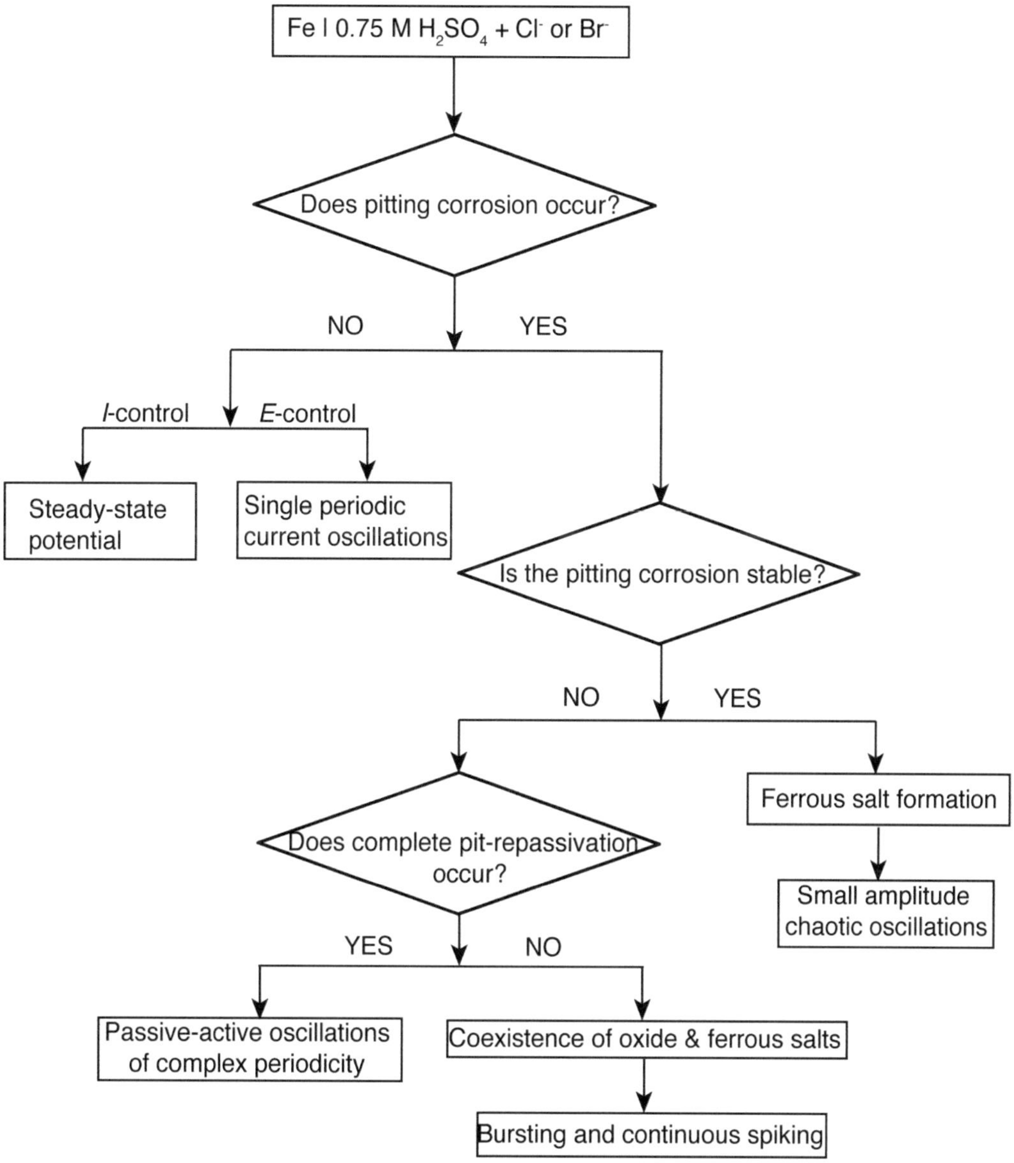

Fig. 15. Flow diagram displaying a phenomenological classification of the nonlinear dynamical response of the halide-perturbed Fe | 0.75 M H_2SO_4 system arisen at various stages of pitting corrosion.

Perturbation with relatively small amounts of Cl^- (c_{Cl^-} <20 mM) leads to unstable pitting corrosion associated with complex passive-active current oscillations arisen within a fixed potential region. These oscillations may be employed to distinguish between general and pitting corrosion and characterize pit initiation and propagation.

In summary, the localized breakdown of passivity on Fe and pit initiation are characterized by:

1. A gradual decrease of the current in the passive state, $I_{pas,f}$ and $I_{pas,b}$ upon increasing gradually the c_{Cl^-}.
2. No access to E_{tr} and the onset of E_{pit}. At potentials higher than E_{pit} a steady pit growth occurs.
3. The disappearance of the single periodic relaxation oscillation of the halide-free system and the onset of complex passive-active oscillations that represent early stages of pitting.
4. The induction time, t_{ind} elapsed before oscillations start. During t_{ind} pit nucleation and repassivation occur repeatedly.
5. Deviation of the $(I_{osc})_{max}$ from the kinetics of the linear part of the active region, which is assigned to the increase of the Fe active surface due to pitting.

Upon increasing c_{Cl^-}> 20 mM and E the rate of pit growth is accelerated resulting in late stages of pitting corrosion. At late stages of pitting corrosion, formation of the iron oxide film becomes unlikely and precipitation of ferrous salts may occur. When a steady pit growth is established and formation of the oxide film becomes unlikely, the precipitation-dissolution of ferrous salt layers results in new oscillatory phenomena related to the following changes:

1. The current in the passive state tends to a limiting current value and a LCR emerges out of the passive state.
2. A critical pitting potential E_{pit} does not exist.
3. At lower potentials either aperiodic bursting oscillations or continuous spiking (beating) of the current (type I) are observed.
4. At higher potentials small amplitude chaotic oscillations (type II) arise around the LCR, instead of the large amplitude oscillations of type I. Beyond a halide concentration threshold, oscillations of type II occur within the entire oscillatory potential region.

Under current-controlled conditions, the nonlinear dynamical response of the halide-perturbed Fe | 0.75 M H_2SO_4 system recorded in E=f(I) and E=f(t) curves is characterized by potential oscillations. It is worth-noting that under a current control the halide free-system exhibits only bistability without oscillations (Fig. 3b). Therefore, potential oscillations can be used alternatively to identify at a first glance pitting corrosion occurring at a critical c_{Cl^-} or c_{Br^-} that depends on the applied current. Iodides do not induce potential oscillations of this type due to the formation of a compact iodine surface layer. Potential oscillations recorded at different c_{Cl^-} or c_{Br^-} exhibit characteristic properties that correspond either to early or late stages of pits. In summary, identification and characterization of pitting corrosion should be based on the following criteria:

1. Onset of potential oscillations in the E = f(I) and E = f(t) curves.
2. The increase of the current I_{act} at which activation of Fe occurs during the backward current scan in the galvanodynamic E = f(I) curves upon increasing gradually either c_{Cl^-} or c_{Br^-}. The I_{act} coincides with the current in the passive state of the potentiodynamic I = f(E) curves and hence quantifies the extent of pitting corrosion and aggressiveness of halides.

3. The decrease of the induction period of time, t_{ind}, elapsed before potential oscillations start, upon increasing gradually either c_{Cl^-} and c_{Br^-} or I_{appl}. The t_{ind} characterizes the kinetics of pit initiation on the passive Fe surface.
4. The decrease of the average firing rate, $<r>$ upon increasing gradually either c_{Cl^-} and c_{Br^-} or I_{appl}. The $<r>$ characterizes pit growth and is associated with the conversion of the outermost oxide layer on the Fe surface to an unstable porous, nonprotective iron chloride or bromide ferrous salt film related rather to electropolishing state dissolution.

The formation of mutually interacted pits on the passive metal surface is necessary for the appearance of potential oscillations. An autocatalytic process formed by a coupling between the oxide detachment and oxide growth causes the repetitive passivation-activation processes resulting in the appearance of the potential oscillation (Sazou et al., 2009).

Analogous phenomena of current and potential oscillations have been also observed for other metals and certainly for several iron alloys during pitting corrosion (Podesta et al., 1979). Thus an approach within the framework of nonlinear dynamics might be used and further developed to study efficiently localized corrosion phenomena in other corroding systems.

8. Conclusions

Pitting corrosion is a complex multi-stage phenomenon of a great technological importance. It has been investigated intensively over many decades. Numerous theoretical and experimental contributions brought about considerable progress in understanding critical factors controlling pitting corrosion. Noticeable progress in elucidating pit nucleation processes during last decades might be attributed to the combined application of electrochemical and surface analytical techniques (Winston Revie, 2011). However, many aspects of pitting corrosion remain unclear.

In this brief review, an alternate route to investigate pitting corrosion is suggested. This includes a closer look on the conditions related to the onset of nonlinear dynamical response of the metal|electrolyte system as well as on characteristics of oscillations related to different stages of pitting. The halide-containing Fe|H_2SO_4 system was selected as a paradigm using current oscillations observed under potentiostatic conditions as well as potential oscillations observed under galvanostatic conditions. Since oscillatory phenomena is a widespread phenomenon in electrochemical reactions, many other metal|electrolyte systems should certainly respond by an oscillatory current and/or potential to a halide ion perturbation. A wide variety of processes can lead to oscillation in the current and potential.

In the case of the halide-perturbed Fe|H_2SO_4 system these processes can be classified at first in two broad categories, those associated with general corrosion and those associated with pitting corrosion. General corrosion corresponds to either a stable steady-state passivity under current- controlled conditions or single periodic current oscillations under potential-controlled conditions. Pitting corrosion corresponds to complex periodic and aperiodic (bursting and continuous spiking) oscillation. Second, processes associated with pitting corrosion can be distinguished to those leading to early stages of pitting and those leading to late stages. At early stages, unstable pitting gives rise to passive-active current and potential oscillation. Both current and potential oscillate between the active state (high current, low potential) and the passive state (low current, high potential). At late stages, the oxide growth becomes unlikely and stable pitting evolves through the precipitation-dissolution of ferrous salt layers whereas complex current or potential oscillations arise. Quantities such as the dissolution current, the induction period required for oscillation to occur and the frequency of oscillations can describe the kinetics of different processes.

It is also worth noting, that current transients of a stochastic nature (noise) of the order of μA might be also induced by halide ions due to randomly nucleated metastable pits over the passive metal surfaces. They occur at potentials lower than the E_{pit}, being the critical potential for pit stabilization. Spatial and temporal interactions among metastable pits leading to clustering and hence high corrosion rates of stainless steel were investigated thoroughly over last decade within the context of nonlinear dynamics and pattern formation (Lunt et al., 2002; Mikhailov et al., 2009; Organ et al., 2005; Punckt et al., 2004). The potential region where these current transients appear is distinctly different from the potential region within which the large-amplitude complex passive-active current oscillations, discussed in this article, arise. Both stochastic noise and deterministic oscillations can be useful in investigating localized corrosion, which is by itself a typical intrinsically complex system (Aogaki, 1999).

This brief review, not necessarily comprehensive, has focused on results from a research project being carried out by our research group over last two decades. It is noticeable that complex and chaotic current or potential oscillations can be further analyzed using numerical diagnostics (i.e. power spectral density, phase portraits, correlation dimension of chaotic attractors, Lyapunov exponents) developed to characterize time series in nonlinear dynamical systems (Corcoran & Sieradzki, 1992; Hudson & Basset, 1991; Kantz & Schreiber, 1997; Karantonis & Pagitsas, 1996; Li et al., 2005; Li et al., 1993). This analysis might provide new diagnostic criteria that can be profitably used in pitting corrosion studies. However, the purpose of this chapter was restricted to point out the rich dynamical response that may arise under appropriate conditions when localized breakdown of the passivity on a metal occurs. It seems that the need for continuing research into the field remains mandatory. It is our belief that the rich nonlinear dynamical response of corrosive systems can be used profitably to gain a further understanding of complex, not-fully understood processes underlying technologically important problems.

9. References

Aogaki, R., Nonequilibrium fluctuations in the corrosion process. In: E. White, et al., (Eds.), *Modern Aspects of Eelctrochemistry*. Plenum Publishers, N.Y., 1999, Vol. 33, pp. 217-305.

Berthier, F., Diard, J.-P., Gorrec, B. L., Montella, C. (2004) Study of the forced Ni | 1 M H_2SO_4 oscillator. *J. Electroanal. Chem.*, 572, 267-281.

Birzu, A., Green, B. J., Jaeger, N. I., Hudson, J. L. (2001) Spatiotemporal patterns during electrodissolution of a metal ring:three-dimensional simulations. *J. Electroanal. Chem.*, 504, 126-136.

Birzu, A., Green, B. J., Otterstedt, R. D., Jaeger, N. I., Hudson, J. L. (2000) Modelling of spatiotemporal patterns during metal electrodissolution in a cell with a point reference electrode. *Phys. Chem. Chem. Phys*, 2, 2715-2724.

Bohni, H., Localized corrosion. In: F. Mansfeld, (Ed.), *Corrosion Mechanisms*. Marcel Dekker, NY, 1987, pp. 285-328.

Budiansky, N. D., Hudson, J. L., Scully, J. R. (2004) Origins of persistent interaction among localized corrosion sites on stainless steel. *J. Electrochem. Soc.*, 151, B233-B243.

Corcoran, S. G., Sieradzki, K. (1992) Chaos during the growth of an artificial pit. *J. Electrochem. Soc.*, 139, 1568-1573.

Dayan, P., Abbott, B., 2001. *Theoretical Neuroscience, Computational and Mathematical Modeling of Neural Systems*. The MIT Press, Cambridge, Massachusetts.

Eiswirth, M., Lubke, M., Krischer, K., Wolf, W., Hudson, J. L., Ertl, G. (1992) Structural effects on the dynamics of an electrocatalytic oscillator. *Chem. Phys. Lett.*, 192, 254-258.

Engell, H. J. (1977) Stability and breakdown phenomena of passivating films. *Electrochim. Acta*, 22, 987-993.

Ertl, G. (1998) Pattern formation at electrode surfaces. *Electrochim. Acta*, 43, 2743-2750.

Ertl, G. (2008) Reactions at surfaces: from atoms to complexity (Nobel lectrure). *Angew. Chem. Int. Ed.*, 47, 3524-3535.

Fechner, G. T. (1828) Zur Elektrochemie-I. Uber Umkehrungen der Polaritat in der einfachen Kette. *Schwigg J Chem. Phys.* , 53, 129.

Franck, U. F. (1978) Chemical oscillations. *Angew. Chem. Int. Ed.* , 17, 1-15.

Franck, U. F., Fitzhugh, R. (1961) Periodische Elektrodenprozesse Und Ihre Beschreibung Durch Ein Mathematisches Modell. *Z. Elektrochemie*, 65, 156-168.

Frankel, G. S. (1998) Pitting corrosion of metals. A review of the critical factors. *J. Electrochem. Soc.*, 145, 2186-2198.

Fujioka, E., Nishihara, H., Aramaki, K. (1996) The inhibition of pit nucleation and growth on the passive surface of iron in a borate buffer solution containing Cl- by oxidizing inhibitors. *Corros. Sci.*, 38, 1915-1933.

Gabrielli, C., Keddam, M., Stupnisek-Lisac, E., Takenouti, H. (1976) Etude du comportment anodique de interface fer-acid nitrique a l'aide d'une regulation a resistance negative *Electrochim. Acta*, 21, 757-766.

Georgolios, C., Sazou, D. (1998) On the mechanism initiating bursting oscillatory patterns during the pitting corrosion of a passive rotating iron-disc electrode in halide-containing sulphuric acid solutions. *J. Solid State Electrochem.*, 2, 340-346.

Geraldo, A. B., Barcia, O. E., Mattos, O. R., Huet, F., Tribollet, B. (1998) New results concerning the oscillations observed for the system iron-sulphuric acid. *Electrochim. Acta*, 44, 455-465.

Green, B. J., Hudson, J. L. (2001) Spatiotemporal patterns and symmetry breaking on a ring electrode. *Phys. Rev. E*, 63, 0226214-1-0226214-8.

Hudson, J. L., Basset, M. R. (1991) Oscilaltory eelctrodissolution of metals. *Rev. Chem. Eng.*, 7, 102-109.

Hudson, J. L., Krischer, J. T., Kevrekidis, I. G. (1993) Spatiotemporal period doubling during the electrodissolution of iron. *Phys. Lett. A*, 179, 355-363.

Hudson, J. L., Tsotsis, T. T. (1994) Electrochemical reaction dynamics. A review. *Chem. Eng. Sci.*, 49, 1493-1572.

Janik-Czachor, M. (1981) An assessment of the processes leading to pit nucleation on iron. *J. Electrochem. Soc.*, 128, 513C-519C.

Kaesche, H., 1986. *Metallic Corrosion*. NACE, TX.

Kantz, H., Schreiber, T., 1997. *Nonlinear Time Series Analysis*. Cambridge University Press, Cambridge, U.K.

Karantonis, A., Pagitsas, M. (1996) Comparative study for the calculation of the Lyapunov spectrum from nonlinear experiemntal signals. *Phys. Rev. E*, 53, 5428-5444.

Karantonis, A., Pagitsas, M. (1997) Constructing normal forms from experimental observations and time series analysis. *Int. J. Bif. Chaos*, 7, 107-127.

Karantonis, A., Pagitsas, M., Miyakita, Y., Nakabayashi, H. (2005) Manipulation of spatio-temporal patterns in networks of relaxation electrochemical oscillators. *Electrochim. Acta*, 50, 5056-5064.

Karantonis, A., Shiomi, Y., Nakabayashi, S. (2000) Coherence and coupling during oscillatory metal electrodissolution. *J. Electroanal. Chem.*, 493, 57-67.

Keddam, M., Lizee, J. F., Pallotta, C., Takenouti, H. (1984) Electrochemical behavior of passive iron in acid medium. 1. Impedance approach. *J. Electrochem. Society*, 131, 2016-2024.

Kiss, I. Z., Hudson, J. L. (2003) Chaotic cluster itirenancy and hierarchical cluster trees in electrochemical experiments. *Chaos*, 13, 999-1009.

Kiss, I. Z., Hudson, J. L., Santos, G. J. E., Parmananda, P. (2003) Experiments on coherence resonance: Noisy precursors to Hopf bifurcations. *Phys. Rev. E*, 67, 035201-035204.

Kiss, I. Z., Lv, Q., Organ, L., Hudson, J. L. (2006) Electrochemical bursting oscillations on a high-dimensional slow subsystem. *Phys. Chem. Chem. Phys.*, 8, 2707-2715.

Kleinke, M. U. (1995) Chaotic behavior of current oscillations during iron electrodissolution in sulfuric acid. *J. Phys. Chem. B*, 99, 17403-17409.

Koper, M. T. M., Oscillation and complex dynamical bifurcation in electrochemical systems. In: I. Prigogine, S. A. Rice, (Eds.), *Adv. Chem. Phys.* Wiley, N.Y., 1996a, Vol. XCII, pp. 161-298.

Koper, M. T. M. (1996b) Stability study and categorization of electrochemical oscillators by impedance spectroscopy. *J. Electroanal. Chem.*, 409, 175-182.

Koper, M. T. M., Sluyters, J. H. (1993a) A mathematical model for current oscillations at the active-passive transition in metal electrodissolution. *J. Electroanal. Chem.*, 347, 31-48.

Koper, M. T. M., Sluyters, J. H. (1993b) On the mathematical unification of a class of electrochemical oscillators and their design procedures. *J. Electroanal. Chem.*, 352, 51-64.

Koutsaftis, D., Karantonis, A., Pagitsas, M., Kouloumbi, N. (2007) Transient and persistent electrochemical bursting induced by halide ions. *J. Phys. Chem. C*, 111, 13579-13585.

Krischer, K., Principles of temporal and spatial formation in electrochemical systems. In: R. E. White, et al., (Eds.), *Modern Aspects of Electrochemistry*. Kluwer Academic/Plenum Publishers, NY, 1999, Vol. 32, pp. 1-142.

Krischer, K., Nonlinear dynamics in electrochemical systems. In: R. C. Alkire, D. M. Kolb, (Eds.), *Advances in Electrochemical Sciences and Engineering*. Wiley-VCH Verlag GMbH & Co. KgaA, Weinheim, 2003b, Vol. 8, pp. 89-208.

Lev, O., Wolffberg, A., Pismen, L. M. (1988) Bifurcations to periodic and chaotic motions in anodic nickel dissolution. *Chem. Eng. Sci.*, 43, 1339-1353.

Li, L., Luo, J. L., Lu, B. T., Chen, S. H. (2005) Effect of interface chloride ion perturbation on oscillatory electrodissolution. *Electrochim. Acta*, 50, 3524-3535.

Li, W. H., Nobe, K. (1993) Electrodissolution kinetics of iron in chloride solutions .9. Effect of benzotriazole on potential oscillations. *J. Electrochem. Soc.*, 140, 1642-1650.

Li, W. H., Nobe, K., Pearlstein, A. J. (1993) Electrodissolution kinetics of iron in chloride solutions .VIII. Chaos in potential/current oscillations. *J. Electrochem. Soc.*, 140, 721-728.

Li, W. H., Wang, X. L., Nobe, K. (1990) Electrodissolution kinetics of iron in chloride solution. 7. Experimental potential/current oscillations. *J. Electrochem. Soc.*, 137, 1184-1188.

Lunt, T. T., Scully, J. R., Brusamarello, V., Mikhailov, A. S., Hudson, J. L. (2002) Spatial interactions among localized corrosion sites. Experiments and modeling. *J. Electrochem. Soc.*, 2002, B163-B173.

Ma, H.-Y., Yin, B.-S., Li, G.-Y., Guo, W.-J., Chen, S.-H., Tang, K. (2003) A study of the relation between current oscillations and pitting. *Chin. J. Chem.*, 21, 1309-1314.

Ma, L., Vitt, J. E. (1999) Current oscillations during iodide oxidation at a gold rotating disk electrode. *J. Electrochem. Soc.*, 146, 4152-4157.

Macdonald, D. D. (1992) The point defect model for the passive state. *J. Electrochem. Soc.*, 139, 3434-3449.

Maurice, V., Marcus, P., Scanning tunneling microscopy and atomic force microscopy. In: P. Marcus, F. Mansfeld, (Eds.), *Analytical Methods in Corrosion Science and Engineering*. CRC Press, Taylor & Francis Group, N.Y., 2006, pp. 133-168.

Mikhailov, A. S., Scully, J. R., Hudson, J. L. (2009) Nonequilibrium collective phenomena in the onset of pitting corrosion. *Surf. Sci.*, 603, 1912-1921.

Nakanishi, S., Sakai, S., Nagai, T., Nakato, Y. (2005) Macroscopically uniform nanoperiod alloy multilayers formed by coupling of electrodeposition with current oscillations. *J. Phys. Chem. B*, 109, 1750-1755.

Newmann, R. C., Ajjawi, M. A. A. (1986) A micro-electrode study of the nitrate effect on pitting of stainless steel. *Corros. Sci.*, 26, 1057-1063.

Organ, L., Sculli, J., Mikhailov, A. S., Hudson, J. L. (2005) A spatiotemporal model of interactions among metastable pits and the transition to pitting corrosion. *Electrochim. Acta*, 51, 225-241.

Orlik, M. (2009) Self-organization in nonlinear dynamical systems and its relation to the materials science. *J. Solid State Electrochem.*, 13, 245-261.

Otterstedt, R. D., Plath, P. J., Jaeger, N. I. (1996) Modulated electrochemical waves. *Phys. Rev. E*, 54, 3744-3751.

Pagitsas, M., Diamantopoulou, A., Sazou, D. (2001) Distinction between general and pitting corrosion based on the nonlinear dynamical response of passive iron surfaces perturbed chemically by halides. *Electrochem. Commun.*, 3, 330-335.

Pagitsas, M., Diamantopoulou, A., Sazou, D. (2002) General and pitting corrosion deduced from current oscillations in the passive-active transition state of the Fe|H_2SO_4 electrochemical system. *Electrochim. Acta*, 47, 4163-4179.

Pagitsas, M., Diamantopoulou, A., Sazou, D. (2003) A point defect model for the general and pitting corrosion on iron|oxide|electrolyte interface deduced from current oscillations. *Chaos Solit. & Fractals*, 17, 263-275.

Pagitsas, M., Pavlidou, M., Papadopoulou, S., Sazou, D. (2007) Chlorates induce pitting corrosion of iron in sulfuric acid solutions: An analysis based on current oscillations and a point defect model. *Chem. Phys. Lett.*, 434, 63-67.

Pagitsas, M., Pavlidou, M., Sazou, D. (2008) Localized passivity breakdown of iron in chlorate- and perchlorate-containing sulphuric acid solutions: A study based on current oscillations and a point defect model. *Electrochim. Acta*, 53, 4784-4795.

Pagitsas, M., Sazou, D. (1991) The improved Franck-Fitzhugh model for the electrodissolution of iron in sulfuric-acid-solutions. Linear stability and bifurcation analysis. Derivation of the kinetic equations for the forced Franck-Fitzhugh model. *Electrochim. Acta*, 36, 1301-1308.

Pagitsas, M., Sazou, D. (1999) Current oscillations induced by chlorides during the passive-active transition of iron in a sulfuric acid solution. *J. Electroanal. Chem.*, 471, 132-145.

Parmananda, P., Rivera, M., Madrigal, R. (1999) Altering oscillatory dynamics of an electrochemical system using external forcing. *Electrochim. Acta*, 44, 4677-4683.

Parmananda, P., Rivera, M., Madrigal, R., Kiss, I. Z., Gaspar, V. (2000) Resonant control of electrochemical oscillations. *J. Phys. Chem. B*, 104, 11748-11751.

Pickering, H. W. (1989) The significance of the local electrode potential within pits, crevices and cracks. *Corros. Sci.*, 29, 325-341.

Pickering, H. W., Frankenthal, R. P. (1972) On the mechanism of localized corrosion of iron and stainless steel. I. Electrochemical studies. *J. Electrochem. Soc.*, 119, 1297-1303.

Podesta, J. J., Piatti, R. C. V., Arvia, A. J. (1979) Potentiostatic current oscillations at iron/sulphuric acid solution interfaces. *J. Electrochem. Soc.*, 126, 1363-1367.

Postlethwaite, J., Kell, A. (1972) Periodic phenomena during the anodic dissolution of iron in sodium chloride solutions. *J. Electrochem. Soc.*, 119, 1351-1352.

Punckt, C., Bolscher, M., Rotermund, H. H., Mikhailov, A. S., Organ, L., Budiansky, N. D., Scully, J. R., Hudson, J. L. (2004) Sudden onset of pitting corrosion on stainless steel as a critical phenomenon. *Science*, 305, 1133-1135.

Rius, A., Lizarbe, R. (1962) Study of the anodic behavior of iron at high potentials in solutions containing chloride ions. *Electrochim. Acta*, 7, 513-422.

Rush, B., Newman, J. (1995) Periodic behavior in the iron sulfuric-acid system. *J. Electrochem. Soc.*, 142, 3770-3779.

Saitou, M., Fukuoka, Y. (2004) An experimental study on stripe pattern formation of Ag-Sb electrodeposits. *J. Phys. Chem. B*, 108, 5380-5385.

Sato, N. (1982) Anodic breakdown of passive films on metals. *J. Electrochem. Soc.*, 129, 255-260.

Sato, N. (1987) Some concepts of corrosion fundamentals. *Corros. Sci.*, 27, 421-433.

Sato, N. (1989) Toward a more fundamental understanding of corrosion processes. *Corrosion*, 45, 354-368.

Sato, N. (1990) An overview on the passivity of metals. *Corros. Sci.*, 31, 1-19.

Sazou, D., Diamantopoulou, A., Pagitsas, M. (2000a) Chemical perturbation of the passive-active transition state of Fe in a sulfuric acid solution by adding halide ions. Current oscillations and stability of the iron oxide film. *Electrochim. Acta*, 45, 2753-2769.

Sazou, D., Diamantopoulou, A., Pagitsas, M. (2000b) Complex periodic and chaotic current oscillations related to different states of the localized corrosion of iron in chloride-containing sulfuric acid solutions. *J. Electroanal. Chem.*, 489, 1-16.

Sazou, D., Diamantopoulou, A., Pagitsas, M. (2000c) Conditions for the onset of current oscillations at the limiting current of the iron electrodissolution in sulfuric acid solutions. *Russ. J. Electrochem.*, 36, 1072-1084.

Sazou, D., Karantonis, A., Pagitsas, M. (1993a) Generalized Hopf, saddle-node infinite period bifurcations and excitability during the electrodissolution and passivation of iron in a sulfuric acid solution. *Int. J. Bif. Chaos*, 3, 981-997.

Sazou, D., Pagitsas, M. (2002) Nitrate ion effect on the passive film breakdown and current oscillations at iron surfaces polarized in chloride-containing sulfuric acid solutions. *Electrochim. Acta*, 47, 1567-1578.

Sazou, D., Pagitsas, M. (2003a) Electrochemical current oscillations during localized corrosion of iron. *Fluct. & Noise Lett.*, 3, L433-L454.

Sazou, D., Pagitsas, M. (2003b) Non-linear dynamics of the passivity breakdown of iron in acidic solutions. *Chaos Solit. & Fractals*, 17, 505-522.

Sazou, D., Pagitsas, M. (2006a) Electrochemical instabilities due to pitting corrosion of iron. *Russ. J. of Electrochem.*, 42, 476-490.

Sazou, D., Pagitsas, M. (2006b) On the onset of current oscillations at the limiting current region emerged during iron electrodissolution in sulfuric acid solutions. *Electrochim. Acta*, 51, 6281-6296.

Sazou, D., Pagitsas, M., Geogolios, C. (1993b) Bursting and beating current oscillatory phenomena induced by chloride-ions during corrosion/passivation of iron in sulfuric acid solutions. *Electrochim. Acta*, 38, 2321-2332.

Sazou, D., Pagitsas, M., Georgolios, C. (1992) The influence of chloride-ions on the dynamic characteristics observed at the transition between corrosion and passivation states of an iron electrode in sulfuric acid solutions. *Electrochim. Acta*, 37, 2067-2076.

Sazou, D., Pavlidou, E., Pagitsas, M. (2011) Potential oscillations induced by localized breakdown of the passivity on Fe in halide-containing sulphuric acid media as a probe for a comparative study of the halide ion effect. *submitted*.

Sazou, D., Pavlidou, M., Pagitsas, M. (2009) Temporal patterning of the potential induced by localized corrosion of iron passivity in acid media. Growth and breakdown of the oxide film described in terms of a point defect model. *Phys. Chem. Chem. Phys.*, 11, 8841-8854.

Schmuki, P. (2002) From Bacon to barriers: a review on the passivity of metals and alloys. *J. Solid State Electrochem.*, 6, 145-164.

Schultze, J. W., Lohrenger, M. M. (2000) Stability, reactivity and breakdown of passive films. Problems of recent and future research. *Electrochim. Acta*, 45, 2499-2513.

Strehblow, H. H., Mechanisms of pitting corrosion. In: P. Marcus, J. Oudar, (Eds.), *Corrosion Mechanisms in Theory and Practice*. Marcel Dekker, NY, 1995, pp. 201-237.

Strehblow, H. H., Wenners, J. (1977) Investigation of processes on iron and nickel electrodes at high corrosion current densities in solutions of high chloride content. *Electrochim. Acta*, 22, 421-427.

Taveira, L. V., Macak, J. M., Sirotna, K., Dick, L. F. P., Schmuki, P. (2006) Voltage oscillations and morphology during the galvanostatic formation of self-organized TiO_2 nanotubes. *J. Electrochem. Soc.*, 153, B137-B143.

Toney, M. F., Davenport, A. J., Oblonsky, L. J., Ryan, M. P., Vitus, C. M. (1997) Atomic structure of the passive oxide film formed on iron. *Phys. Rev. Lett.*, 79, 4282-4285.

Vetter, K. J. (1971) General kinetics of passive layers on metals. *Electrochim. Acta*, 16, 1923-1937.

Vitt, J. E., Johnson, D. C. (1992) The importance of anodic discharge of H_2O in anodic oxygen-transfer reactions. *J. Electrochem. Soc.*, 139, 774-778.

Wang, C., Chen, S. (1998) Holographic microphotography study of periodic electrodissolution of iron in a magnetic field. *Electrochim. Acta*, 43, 2225-2232.

Wang, C., Chen, S., Yang, X., Li, L. (2004) Investigation of chloride-induced pitting processes of iron in the H_2SO_4 solution by the digital holography. *Electrochem. Commun.*, 6, 1009-1015.

Winston Revie, R., 2011. *Uhlig's Corrosion Handbook*. John Wiley & Sons, Inc. , N.J., U.S.A.

5

Systemic and Local Tissue Response to Titanium Corrosion

Daniel Olmedo[1,2], Deborah Tasat[1,3], Gustavo Duffó[2,3,4], Rómulo Cabrini[1,4] and María Guglielmotti[1,2]
[1]*University of Buenos Aires*
[2]*National Research Council (CONICET)*
[3]*National University of General San Martin*
[4]*National Atomic Energy Commission*
Argentina

1. Introduction

The term biomaterials refers to materials that have been designed to be implanted or placed inside a live system with the aims to substitute or regenerate tissue and tissue functions. Williams defines biomaterials as those that are used in devices for biomedical use designed to interact with biological systems (Williams, 1986). Classically, biomaterials are divided into four types: polymers, metals, ceramics and natural materials. Two different types of biomaterials can be combined to obtain a fifth type known as composite biomaterials (Abramson et al., 2004). Biomaterials are widely used in orthopedic, dental, cardiovascular, ophthalmological, and reconstructive surgery, among other applications. The discovery of relatively inert metals and alloys has led to their use in the field of biomedical applications such as orthopedics and dentistry, and their use in increasing due to their physical-chemical properties and compatibility with biological surroundings (Ratner et al., 2004). One of the most frequently employed metallic biomaterials is titanium (Anderson et al., 2004). Though zirconium is not widely used as a clinical material, it is chemically closely related to and has several properties in common with titanium (Thomsen et al., 1997). Although both titanium and zirconium are transition metals, their physicochemical properties such as oxidation velocity, interaction with water, crystalline structure, transport properties, and those of their oxides differ quantitatively (Henrich & Cox, 1994); these differences may have an effect on biological response (Thomsen et al., 1997). Indeed, the use of zirconium and zirconium alloys to manufacture implants for traumatological, orthopedic, and dental applications has been reported (Sherepo et al., 2004; Sollazzo et al., 2007).

Titanium and zirconium are highly reactive metals and when exposed to fluid media or air, they quickly develop a layer of titanium dioxide (TiO_2) or zirconium dioxide (ZrO_2). This layer of dioxide forms a boundary at the interface between the biological medium and the metal structure. It produces passivation of the metal, determining the degree of biocompatibility and the biological response to the implant (Kasemo 1983, Kasemo & Lausmaa 1988, Long & Rack, 1998). Titanium dioxide exists naturally, mainly in the form of three crystalline structures: rutile, anatasa, and brookite. In the case of titanium implants, the passive oxide layer is composed of anatase and rutile or anatase alone (Effah et al., 1995; Olmedo et al., 2008a; Sul et al., 2001). Zirconium, however, does not exist as a free metal in nature; it occurs as the minerals zircon, or zirconium silicate ($ZrSiO_4$), and the rare mineral baddeleyite or zirconium dioxide (ZrO_2) which has a monoclinic crystal structure (Zirconium. Mineral Information Institute, 2009). Baddeleyite, also known as zirconia, is the most naturally occurring form and can be transformed into a tetragonal (1100 °C) or cubic (2370 °C) crystallographic form depending on temperature (Chowdhury et al., 2007; Manicone et al., 2007).

Titanium is widely used in the manufacture of dental and orthopedic implants due to its excellent biocompatibility. The latter is defined as the ability of a material to perform with an appropriate host response in a specific application (Williams, 1987). The use of titanium dental implants has revolutionized oral implantology. Currently, almost 300,000 patients in the United States have dental implants. In the area of orthopedics, replacement hip joints are implanted in more than 200,000 humans each year (Ratner et al., 2004). Dental implants are surgically inserted into the jaw bone primarily as a prosthetic foundation. The process of integration of titanium with bone was termed "osseointegration" by Brånemark (Brånemark et al., 1977; Chaturvedi, 2009).

No metal or metal alloy is completely inert *in vivo*. Corrosion is the deterioration of a metal due to interaction (electrochemical attack) with its environment, which results in the release of ions into the surrounding microenvironment (Jacobs, 1998). There are "noble" metals such as rhodium (Rd), palladium (Pd), iridium (Ir) and platinum (Pt), whose resistance to corrosion is due to their high thermodynamic stability. Passivating metals, such as titanium (Ti), vanadium (V), zirconium (Zr), niobium (Nb), and tantalum (Ta), however, are thermodynamically unstable and their resistance to corrosion results from the formation of a protective oxide layer on their surface (Lucas et al, 1992). Titanium is available as commercially pure (c.p.) titanium or as Ti-6Al-4V alloy with 6% aluminum and 4% vanadium. The addition of Al and V increases strength and fatigue resistance; however, this may affect the corrosion resistance properties and may result in the release of metal ions (Textor et al., 2001). C.p. titanium and Ti-6Al-4V alloy are the two most common titanium-based implant biomaterials (Abramson et al., 2004). There are four standard types or grades of c.p. titanium used for the manufacture of surgical implants, which differ in their content of interstitial elements. This content determines the mechanical properties of a material: the higher the content the higher the grade. In other words, grade 1 is the most pure and grade 4 contains the greatest amount of impurities and has the greatest mechanical resistance. C.p. titanium is used to manufacture dental implants, whereas a Ti6Al4V alloy is used mostly in orthopedics.

As previously stated, all the metallic materials employed in surgery as permanent implants are liable, to a certain degree, to corrosion due to variations in the internal electrolyte milieu (Jacobs, 1998). Corrosion, one of the possible causes of implant failure, implies the dissolution of the protective oxide layer. When metal particles/ions are released from the implant surface, they can migrate systemically, remain in the intercellular spaces near the site where they were released, or be taken up by macrophages (Olmedo 2003, 2008b). The presence of metallic particles in peri-implant tissues may not only be due to a process of electrochemical corrosion but also to frictional wear, or a synergistic combination of the two.

Moreover, mechanical disruption during insertion, abutment connection, or removal of failing implants has been suggested as a possible cause of the release of particles from metal structures (Flatebø, 2006; Jacobs, 1998). The release of particles/ions from the implant into the surrounding biological compartment, their biodistribution in the body, and their final destination are issues that lie at the center of studies on biocompatibility and biokinetics. The chemical forms of these released elements have not been identified to date. It is unclear whether these products remain as metal ions or metal oxides, or whether they form protein or cell-bound complexes (Brown et al., 1987; Urban et al., 2000). In the particular case of titanium, little is known about the valence with which it exerts its action, the organic or inorganic nature of its ligands, and its potential toxicity (Jacobs, 1991).

The potential toxicity and biological risks associated with ions and/or particles released due to corrosion of metallic implants is a public health concern for the community of patients who have a prosthesis (orthopedic and/or dental), since these prostheses remain inside the body over long periods of time. Likewise, the subject of corrosion is of interest to researchers; corrosion studies aim at avoiding the possible corrosion-related health problems that may arise when metallic implants are placed in humans. Controlling corrosion is most relevant for, in order to protect patient health, corrosion should be negligible. Thus, managing and controlling corrosion of a biomedical implant is a paramount issue from a biological, sanitary, metallurgic, economic and social point of view. The current massive use of these metal biomaterials in the biomedical field renders it necessary to have detailed knowledge not only on their early effects (short term failure) but especially on their long term effects, given that these materials remain inside the patients over long periods of time, sometimes throughout their entire life. With the aims to improve biocompatibility and mechanical resistance, manufacturers of biomedical implants seek to develop an adequate design with minimal degradation, corrosion, dissolution, deformation, and fracture.

The study of corrosion requires an interdisciplinary approach including chemists, biologists, physicists, engineers, metallurgists, and specialists in biomedicine. The Biomaterials Laboratory of the Department of Oral Pathology of the University of Buenos Aires, the National Commission of Atomic Energy and the University of San Martin have been conducting collaborative research on corrosion aimed at evaluating both local tissue response in the peri-implant microenvironment and the systemic effects and possible consequences of corrosion, focusing mainly on dental implants (Olmedo et al., 2009).

2. Local effects of corrosion

As mentioned above, the titanium dioxide layer prevents corrosion. However, this layer is prone to break, releasing ions/ particles into the milieu. The potential risk of corrosion and the detrimental consequences of corrosion byproducts in the surrounding tissue are issues of clinical importance (Kumazawa, 2002).

The Biomaterials Department has a Failed Human Dental Implants Service devoted to the *in situ* evaluation of the interface, which consists of the implant and the peri-implant tissues, using systematic histological studies. Studying the implant-tissue interface allows detecting osseointegration, implant- marrow tissue interface (myelointegration), fibrous tissue (fibrointegration) and/or inflammatory reactions. According to the experience of our laboratory, the most frequent causes of dental implant failure in humans are mobility, fracture of the metal (fatigue), and early exposure (Guglielmotti & Cabrini, 1997). Interestingly, implants that had failed due to metal fatigue were found to show satisfactory osseointegration, in other words, good integration of titanium with bone. This means the implants were successful from a biological point of view (osseointegration) but a clinical failure from a mechanical viewpoint. Studying the peri-implant tissue at the metal/tissue level allows obtaining relevant data to determine the possible cause/causes of implant failure.

2.1 Tissue response at the metal-tissue interface

Throughout their histologic studies of failed dental implants, Guglielmotti & Cabrini (1997) consistently observed metal particles inside osseointegrated bone tissue and bone marrow of implants that had failed due to metal fatigue, thus finding evidence of corrosion of the metal structure. Similarly, Olmedo et al. (2003a) found macrophages loaded with metal-like particles in peri-implant soft tissues of failed human dental implants indicating the occurrence of corrosion processes (Fig. 1 A-B). Microchemical analysis of the metallic particles inside macrophages using X-ray dispersion (EDX) confirmed the presence of titanium. It is noteworthy that a greater number of macrophages loaded with particles was observed in the vicinity of the metal surface than at more distant sites. Likewise, numerous case reports in the literature describe histological evidence of inflammatory response and the presence of metallic ions/particles in the tissues adjacent to orthopedic prostheses of titanium or titanium based alloys (Jacobs, 1998).

Titanium is widely used in oral and maxillofacial materials such as grids, fixation plates, screws, and distractors. According to a number of studies reported in the literature, the removal of titanium miniplates after bone healing is complete is unnecessary precisely due to the excellent biocompatibility and corrosion resistance properties of titanium. This is beneficial to the patient since a second surgery is avoided (Rosenberg et al., 1993). Moreover, some authors suggest that miniplates must be removed only when they cause patient complaints and in cases of wound dehiscence or infection (Rosenberg et al., 1993). However, as mentioned above, no metal or alloy is completely inert *in vivo*. In this regard, some authors claim that titanium miniplates should be removed to allow for physiologic bony adaptation and avoidance of a foreign body reaction (Ferguson, 1960; Katou et al., 1996; Moran et al., 1991; Rosenberg et al., 1993; Young-Kyun et al., 1997).

Thus, whether titanium miniplates or grids should be removed after bone healing is complete remains controversial to date. Bessho & Iizuka (1993) examined 113 titanium miniplates that had been retrieved after miniplate fixation of mandibular fractures, and identified surface depressions apparently caused by pitting corrosion (Matthew et al., 1996). Zaffe et al. (2003) evaluated the pre and post-implantation surface features and surface alterations of titanium grids and plates in patients and observed, among other alterations, the presence of pitting on the surface of one of the grids. Experimental studies analyzing the biological effect of a type of localized corrosion, pitting corrosion, on the peri-implant environment have been conducted at our laboratory (Olmedo, 2008c). Pitting corrosion

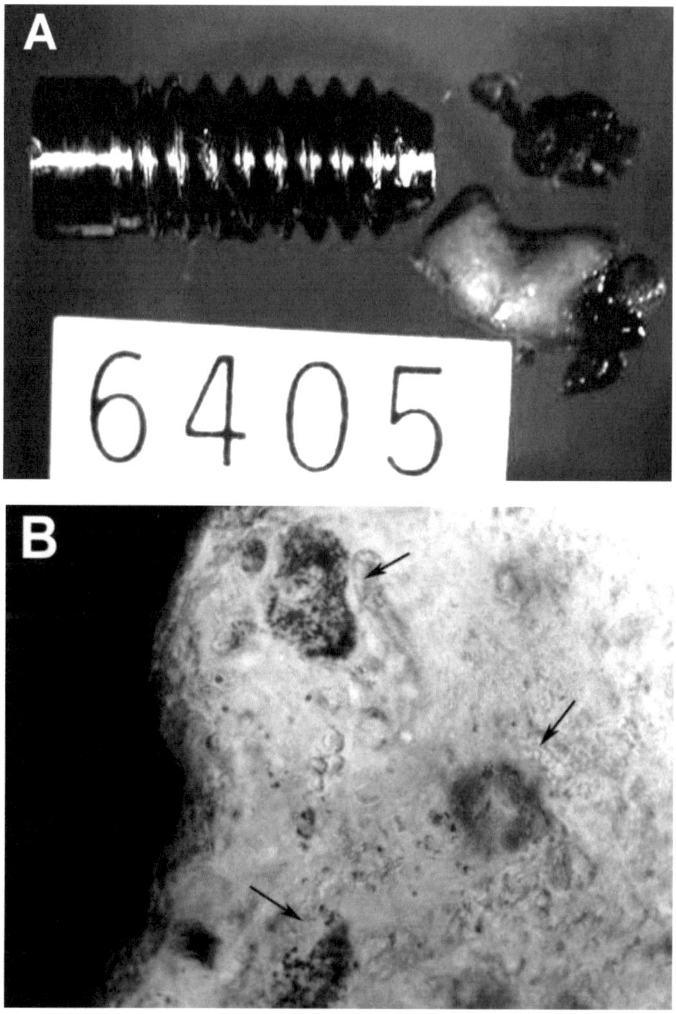

Fig. 1. A) Failed dental implant that shows tissue fragments obtained by curettage of the surgical bed. B) Photomicrograph of macrophages near the surface of the implant (→). Note the presence of particles in their cytoplasm. Ground section. Orig. Mag. X1000

produces local attack, especially on isolated spots of the passivated metal surface, propagating into the metal. The histologic results of our studies showed scarce osseointegration at the bone-implant interface, i.e. the lack of a union between the bone tissue and the surface of the implant; osseointegration was only observed at sites where the metal remained passivated (areas with no pitting and/or surface alterations) (Fig. 2 A,B). The decrease in the percentage of osseointegration in the areas corresponding to the pits would be associated to a change in the chemical composition and/or structure (e.g. crystallography) of the oxide on the pit surface. It is important to point out that the presence of particulate corrosion and wear products in the tissue surrounding the implant may ultimately result in a cascade of events leading to periprosthetic bone loss (Jacobs, 1998; Urban, 1994). The microchemical analysis of corrosion products by energy dispersive x-ray analysis (EDX) in the peri-implant milieu revealed the presence of titanium. It is noteworthy that craters, pits, surface cracks, and depressions may appear during the preparation of the sheets that will be used to manufacture miniplates (Matthew et al., 1996) and may be potential sites for the initiation of corrosion. The results obtained in our study by scanning electron microscopy showed initiation of pitting in areas with surface cracks. Titanium exhibits the greatest resistance to generalized corrosion, pitting corrosion, and crevice corrosion compared to other metals or alloys used in oral surgery, such as stainless steel or chromium-cobalt (Matthew et al., 1996). The severity of corrosion and the quantity of corrosion products that are released may depend not only on the susceptibility to corrosion of the implant material but also on the tissue response to the implant and to the surgical procedures used during implantation (Moberg et al., 1989). The histological results of our study showed the presence of corrosion products around the implant, both outside and phagocytosed in macrophages. In various cases the products of corrosion were found around the blood vessels (Fig. 2 C), in keeping with the histological study of soft tissue adjacent to titanium implants reported by Meachim & Williams (1973) and Torgersen et al. (1995). The observation of metal particles located intracellularly or in association with vessels may represent a biologic response aimed at eliminating the foreign material (Meachim & Williams, 1973; Schliephake et al., 1993; Torgersen et al., 1995).

The properties and quality of the implant material, the shape of the implant, and the handling and surgical procedure are of crucial importance for an optimal biologic performance of any implant device. Unstable conditions in the fracture area after osteosynthesis lead to continuous fretting at the screw/plate interface. Removal of the passivating surface oxide and oxygen depletion in the crevices between plate and screws increase the risk of both crevice corrosion and fretting attack (Williams, 1982). It is speculated that an increased mechanical stability during healing may reduce the fretting component, and thereby reduce corrosion. The adverse local effects caused by pitting corrosion suggest that titanium plates and grids should be used with caution as permanent fixation structures.

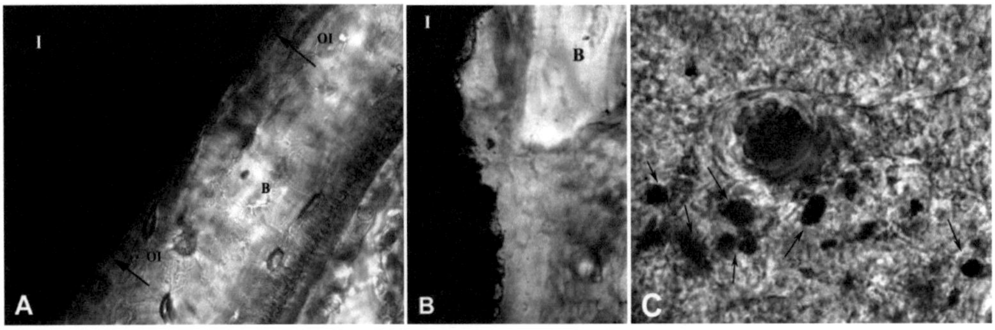

Fig. 2. A and B) Bone tissue-implant interface. A) Control case showing adequate osseointegration (OI) of the bone tissue (B) with the surface of the titanium implant (I). B) Experimental case (pitting corrosion). Note the irregularities on the implant surface (I) and bone tissue (B) far from the surface (lack of osseointegration). C) Blood vessel in the bone marrow near an implant surface and products of corrosion (→) in the vicinity. Ground sections. Orig. Mag. X1000

Based on the aforementioned observations, the occurrence of corrosion phenomena at the interface is of paramount importance to the clinical course of both dental and orthopedic implants since such phenomena could be a possible cause of mid-term implant failure.

2.2 Peri-implant mucosa response

The gingiva around dental implants is called peri-implant mucosa, and consists of well-keratinized oral epithelium, sulcular epithelium, and junctional epithelium with underlying connective tissue. Between the implant surface and epithelial cells are hemidesmosomes and the basal lamina (Newman & Flemming, 1988).

Gingival hyperplasia, mucositis, and peri-implantitis have been described amongst the soft tissue complications associated to dental implants (Adell et al., 1981; Lang et al., 2000). The causes that lead to the development of reactive lesions associated to dental implants have not been fully elucidated to date. In this regard, our research group has reported two clinical cases of reactive lesions in the peri-implant mucosa (inflammatory angiohyperplastic granuloma and peripheral giant cell granuloma) associated to dental implants, in which the presence of metallic particles was detected histologically (Fig. 3 A-B). The presence of metallic particles in the studied tissue suggests that the etiology of the lesions might be attributed to a corrosion process of the metal structure (Olmedo et al., 2010).

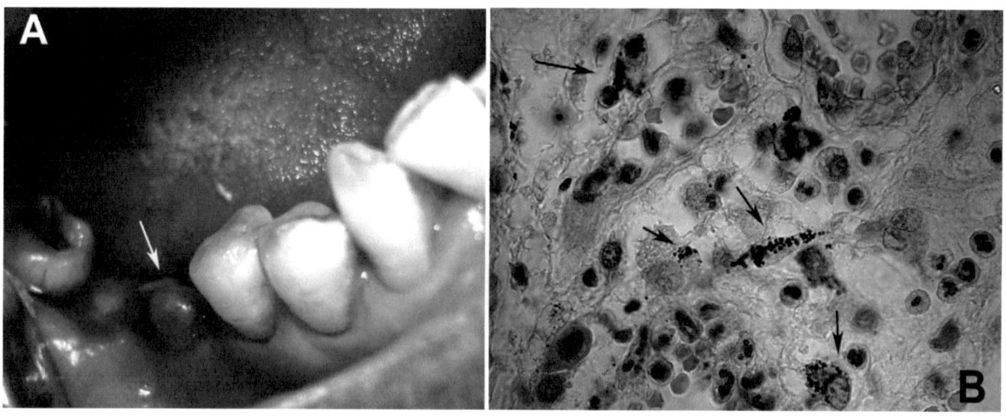

Fig. 3. A) Clinical intraoral photograph showing an exophytic lesion (→) in the area of the first lower right molar. B) Reactive lesion (pyogenic granuloma). Note the significant vascular proliferation and the presence of metal-like particles inside macrophages (→). H-E; Orig. Mag. X400

Abraham et al. (2006) demonstrated the presence of titanium in saliva and gingival fluid of patients carrying titanium dental implants. According to the authors, the highest titanium levels corresponded to patients carrying implants over longer periods of time, thus indicating that titanium accumulates in peri-implant gingival tissue. Oral exfoliative cytology is a diagnostic method which involves the study and interpretation of the features of cells exfoliated from the oral mucosa (Diniz-Freitas et al., 2004). Thus, we performed an exploratory work using exfoliative cytology around the peri-implant mucosa of human dental implants (Nalli et al., 2009). The cytological smears of patients carrying dental implants exhibited metal-like particles varying in quantity, shape, and size. The particles were found both inside and among epithelial cells and macrophages.

The results of the study showed that ions/particles are released from the surface of the implant into the biological milieu. Both epithelial cells and macrophages located in the peri-implant area are able to capture these metal-like particles. Thus, exfoliative cytology is a simple, minimally-invasive, well-tolerated technique, which may prove useful to detect metal particles in cells exfoliated from the peri-implant mucosa, and be a valuable method to monitor dental implant corrosion.

The peri-implant milieu consists of bone tissue, soft tissues, and saliva. Biochemical changes in the peri-implant environment may lead to implant corrosion. According to Abraham et al. (2006) the molecular mechanism of interaction between metal ions and biological molecules or cells remains unclear to date.

The release of ions/particles can cause pigmentation of soft tissues adjacent to an implant (metallosis). Metallosis is defined as aseptic fibrosis, local necrosis, or loosening of a device secondary to metal corrosion and release of wear debris (Black et al., 1990; Bullough, 1994). It involves deposition and build-up of metallic debris in the soft tissues of the body. In a previous study we evaluated histologically tissue response in human oral mucosa associated to submerged titanium implants, using biopsies of the supra-implant oral mucosa adjacent to the implant cover screw (Olmedo et al., 2007a). We observed the presence of different sized particles inside cells or phagocytosed in macrophages in epithelial and connective tissue (Fig. 4 A-B). Interestingly, the titanium particles in the superficial layers of the epithelium might have been associated not only with the cover screw surface but also with other exogenous sources. For example, titanium oxide (TiO_2) is widely used in food products, toothpastes, prophylaxis pastes and abrading and polishing agents, which have been reported in oral biopsies (Koppang et al., 2007).

Microchemical analysis by EDX revealed the presence of titanium in the particles. Immunohistochemical staining with antibodies anti CD68 and anti CD45RO was positive, confirming the presence of macrophages and T lymphocytes associated with the metal particles. In agreement with other reports, (Evrard et al., 2010; Lalor et al., 1991; Matthew et al., 1996;) the T-lymphocyte infiltrate would seem to suggest the presence of an immune response mediated by cells.

Scanning electron microscopy allowed visualizing depressions and irregularities on the surface of the studied metal cover screws. Both unused cover screws and those removed from patients exhibited alterations on their surface. As mentioned previously, craters, pits, surface cracks and depressions may appear during the preparation of the sheets that will be used to manufacture miniplates (Matthew et al., 1996), and be potential sites for the initiation of corrosion. Based on this observation, it would seem advisable for professionals to handle cover screws with utmost care since the observed scratches were most likely caused during placement or removal of the cover screws and could also be potential sites for the initiation of corrosion.

The potential long-term biological effects of particles on soft tissues adjacent to metallic devices should be further investigated as these effects might affect the clinical outcome of the implant.

Corrosion is not only a local problem since the particles released during this process can migrate to distant sites. This issue is of particular interest to biocompatibility studies.

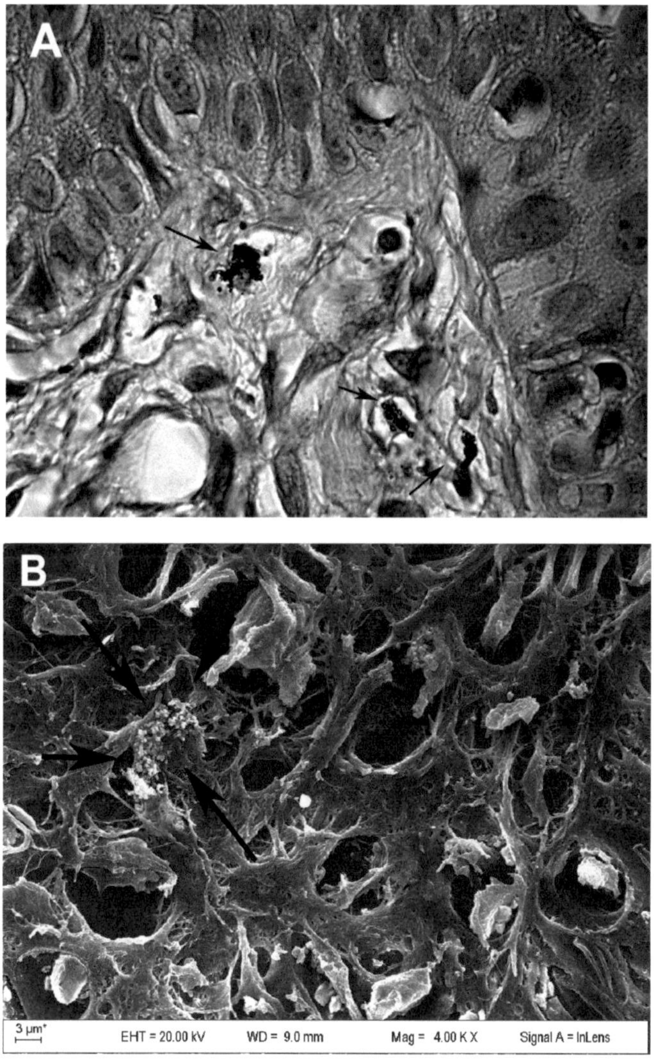

Fig. 4. A) Human oral mucosa covering an implant cover screw. Note the presence of titanium particles (→) inside cells or phagocytosed in macrophages at the epithelium-chorion interface. H-E. Orig. Mag. X1000. B) Scanning Electron Microscopy of an area of mucosa with particles. Note the fine particles (→) among connective tissue elements. Orig. Mag. X4000

3. Studies on the dissemination of titanium towards other biological compartments

The local effect of corrosion and subsequent release of ions/metal-like particles into the peri-implant biological milieu could compromise other biological compartments. The chemically active metal ions/particles may bind to the surrounding tissues but may also bind to proteins and be disseminated in the vascular and lymphatic systems to distant organs (Jacobs et al., 1991; Woodman et al., 1984a).

Studies in the field of orthopedic implants show that titanium ions enter neighboring tissues reaching the internal milieu and are excreted through urine (Jacobs et al., 1991). A number of researchers have found metal ions in body organs and fluids. Jacobs et al. (1991) studied osseointegrated coxofemoral prostheses made of 90% titanium-6%aluminum-4%vanadium and showed that ions of all three metals entered the plasma and were excreted through urine. A study at autopsy by Urban et al (2000) demonstrated the presence of metal-like and plastic particles from coxofemoral and knee-replacement prostheses in the liver, spleen, and lymph nodes.

3.1 Deposition of titanium and zirconium in organs with macrophagic activity. An experimental model

As mentioned previously, titanium and zirconium implants have a protective dioxide (TiO_2 or ZrO_2) layer on their surface. This layer determines biocompatibility and forms a boundary at the interface between the biological milieu and the implant, decreasing their reactivity and partially avoiding corrosion (Jacobs et al., 1998; Kasemo, 1983; Long & Rack, 1998). In order to evaluate the dissemination routes of corrosion products and estimate the intensity of the deposits in different biological compartments, our research group has developed experimental models with animals intraperitoneally injected with TiO_2 or ZrO_2 (Cabrini et al. 2002, 2003; Olmedo et al. 2002, 2003b, 2005, 2008a).

Though it holds true that the experimental doses employed in those studies are high in terms of a normal *in vivo* situation, they served the purpose of our studies since they allow rapid observation of the adverse effects of particles in the studied tissue (Olmedo et al., 2011). Our studies included histologic observation and quantitation of titanium and zirconium deposits in organs with macrophagic activity such as the liver, spleen, and lungs (Fig. 5 A-C), (Olmedo et al., 2002), and showed that at equal doses and experimental times titanium content in organs was consistently higher than zirconium content. Macrophages are cells that respond rapidly to *in vivo* implantation of a biomaterial, including metals, ceramics, cement, and polymers. Their response depends mainly on the size and structure of

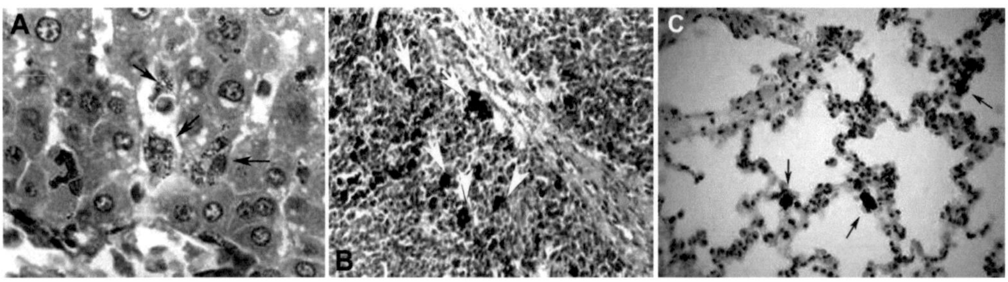

Fig. 5. Titanium deposits in organ parenchyma of an animal injected with TiO_2. A) Liver. Deposition in liver cells (hepatocytes) (→) can be seen clearly. Grenacher carmin stain. Orig. Mag. X400. B and C) Spleen and lung, respectively. Note the amount of titanium (→) deposits. Grenacher carmin stain. Orig. Mag. X400

the material (Anderson et al., 2004; Lu et al., 2002; Solheim et al., 2002; Takebe et al. 2003; Xia & Triffitt, 2006). Particles that are smaller than the macrophages themselves (< 10 μm) can be easily phagocytosed. However, the larger particles (10-100 μm) are ingested by giant, multinucleate cells (Brodbeck et al., 2005). The biokinetics of TiO_2 and ZrO_2 microparticles depends on differences in physicochemical properties of the particles, such as size, shape and/or crystal structure (Olmedo et al., 2011).

3.2 An experimental model

Experimental studies performed at our laboratory showed the presence of titanium and zirconium particles in monocytes in the blood (Fig. 6) and blood plasma (Olmedo et al., 2003b, 2005). Several transport mechanisms have been described for titanium, e.g. systemic dissemination by the vascular system in solution or as particles (Meachim & Williams, 1973); lymphatic dissemination as free particles or as phagocytosed particles within macrophages (Urban et al., 2000), dissemination of particles to the bone marrow by circulating monocytes, or as minute particles by the vascular system (Engh et al., 1997). Several studies on the bond between metal and proteins have contributed to the understanding of the dissemination of metals. Nickel, chromium, and cobalt would seemingly migrate bound to blood cells and/or proteins in serum and tissue fluids (Brown et al., 1987; Merritt et al., 1984). Aluminum is seemingly transported by transferrin (Alfrey, 1989). Uranium is transported linked to proteins, to citrates and to carbonates (Leggett, 1989).

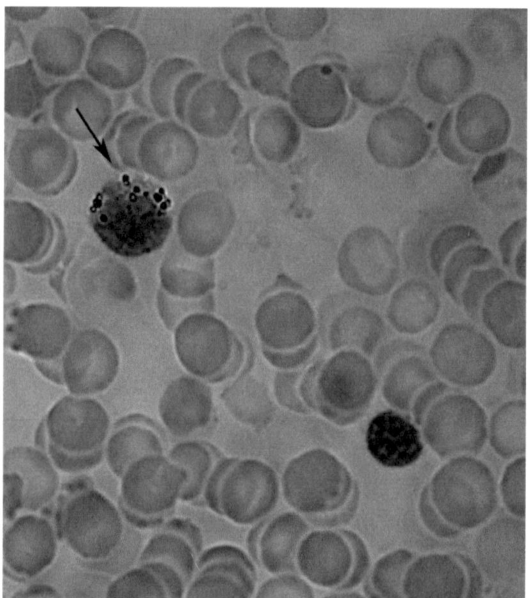

Fig. 6. Blood smear. Titanium particles are evident in a peripheral blood monocyte (→). Safranin stain. Orig. Mag. X1000

The fact that metals bind mainly to albumin would explain their widespread presence in the body. The metallic ions that result from the process of corrosion would thus disseminate to tissues, bind to albumin and enter the circulation exerting their effect at remote sites. Testing for titanium or zirconium in the blood (cells and/or plasma) of patients carrying an implant (coxofemoral implant, dental implant, plates and screws for fracture fixation, metallic panels for reconstructive surgery of large areas of the body) may serve as a method to detect the presence of a corrosion process of the metallic structures (Olmedo et al, 2003b).

3.3 Effect of titanium and zirconium deposition on the lungs: Generation of Superoxide anion (O_2^-) in alveolar macrophages. An experimental model

It is known that trace metals can increase physiological production of reactive oxygen species (ROS) which, without a compensatory increase in antioxidative species, can lead to tissue damage (Gottschling et al., 2002; Kawanishi et al., 2002; Maziere et al., 2003). Studies conducted at our laboratory have shown the presence of titanium and zirconium particles in alveolar phagocytes immunohistochemically identified as CD68 macrophages (Olmedo et al., 2008b). Evaluation of oxidative metabolism of alveolar macrophages exposed to these oxides has shown an increase in generation of ROS. However, it must be pointed out that ROS levels in animals exposed to ZrO_2 were found to be markedly lower than those of animals exposed to TiO_2.

As mentioned previously, the layer of titanium dioxide is crystallographically composed of anatase or a combination of anatase and rutile. Studies on generation of superoxide anion (O_2^-) in alveolar macrophages performed at our laboratory showed that rutile is less bioreactive than anatase. Our results suggest that a rutile coating on metallic biomaterials would improve their biocompatibility properties (Olmedo et al., 2008a).

4. Clinical implications of corrosion

The results of our studies on failed human dental implants and data obtained using the experimental models developed at our laboratory show that any titanium surface can suffer corrosion processes and release particles into the local and systemic biological milieu.

The peri-implant milieu consists of bone tissue, soft tissues, and saliva. Biochemical changes in the peri-implant environment may lead to implant corrosion (Laing, 1973). Thus, titanium implant corrosion is affected not only by the concentration of electrolytes but also by saliva pH (Duffó et al., 1999; Nikolopoulou, 2006) which can vary in areas around dental implants (Meffert et al., 1992). Lotthar et al. (1992) reported that titanium does not withstand a large number of chemical substances. These substances may be in foods, saliva, tooth pastes, and prophylactic agents. They decompose foodstuffs, change plaque metabolism, and cause corrosion (Lotthar et al., 1992; Siirilä & Könönen, 1991). The drop in pH in the electrolytic milieu as a result of local inflammatory processes would seem to stimulate the process of corrosion (Duffó et al., 1999). Abraham et al. (2006) demonstrated the presence of titanium in a wide range of concentrations in saliva and gingival fluid of patients with titanium dental implants.

Significant decreases in pH have been observed in traumatized tissues; indeed pH drops to as low as 4 during the wound healing process (Duffó et al., 1999; Laing, 1973). These low values increase tissue aggressiveness toward the metallic materials. In previous works we found that the decrease in the pH of the electrolytic milieu resulting from local inflammatory processes also stimulates the corrosion process (Duffó et al., 1999).

A corrosion process can decrease the fatigue resistance of the metal compromising metal resistance, which could eventually cause implant fracture (Adya et al., 2005; Guindy et al., 2004 ; Nikolopoulou, 2006 ; Tagger Green et al., 2002). It has been reported that the infiltration of saliva between the suprastructure (nickel-chromium-molybdenum alloy) and the implant (pure titanium) can trigger corrosion processes (galvanic corrosion) due to differences in electrochemical potentials. This causes the release of ions, such as nickel or chromium ions, from the alloy in the crown or bridge to the peri-implant tissues and subsequently results in bone resorption. The latter compromises implant stability, eventually causing implant fracture (Tagger Green et al., 2002).

Metal corrosion can affect the close contact between the implant and the bone tissue (osseointegration). The ions/metallic particles from coxofemoral prostheses can be phagocytosed by macrophages stimulating the release of cytokines, which contribute to bone resorption by activating osteoclasts. In addition to increasing bone resorption, the released particles may inhibit osteoblast function decreasing bone formation and contributing to osteolysis (Allen et al., 1997; Dowd et al., 1995).

The products of metallic implant corrosion behave as haptens generating a hypersensitive reaction that involves the release of inflammatory mediators, known as cytokines, and macrophage recruitment (Hallab et al., 2001; Jiranek et al., 1993; Yang & Merrit, 1994). It remains unclear to date whether it is the hypersensitivity to metal that causes implant failure or vice versa (Hallab et al., 2001). It also remains controversial whether an inflammatory process generates corrosion or whether corrosion triggers an inflammatory reaction. Thus, hypersensitivity to titanium as an implant material in oral and maxillofacial surgery probably occurs more commonly than has been reported in the literature (Matthew & Frame, 1998). There are reports of cases where titanium allergy mainly appeared as the fundamental cause of urticaria, eczema, oedema, redness and pruritus of the skin or mucosa, either localized, at distant sites, or generalized (Sicilia et al, 2008). However, the clinical relevance of allergic reactions in patients with titanium dental implants remains debatable (Javed et al., 2011).

Mineral elements play a critical role in the physiology and pathology of biological systems. Titanium is a nonessential element in that (a) no enzymatic pathway has been elucidated that requires titanium as a cofactor, (b) there does not appear to be any homeostatic control of titanium, and (c) titanium is not invariably detected in the newborn (Woodman et al, 1984b). Thus, the presence of titanium in the body, titanium biokinetics, and the potential biological effects of titanium are of great interest to researchers.

The toxicology of titanium is a current issue of debate. According to epidemiological studies, inhalation of powder containing titanium has no deleterious effect on the lungs (Daum et al., 1977; Ferin & Oberdörster, 1985). Other studies, however, suggest an association between titanium particles and pleural pathologies (Garabrant et al., 1987), granulomatous diseases, and malignant neoplasms of the lung. Our experimental studies have shown the presence of a considerable amount of titanium particles not only in alveolar macrophages but also in hepatocytes (Olmedo et al., 2008b). The accumulation of particles in the liver could compromise liver function as described by Urban et al. (2000). The authors associated the presence of titanium particles in a patient to granulomatous reactions and hepatomegalia. Various studies have reported the presence of macrophages related to failed prostheses, both orthopedic and dental (Adya et al., 2005; Langkamer et al., 1992; Lee et al., 1992; Olmedo et al., 2003; Urban et al., 2002).

As to carcinogenic potential, there are scant reports on the potential development of malignant tumors associated with prosthetic structures in humans (Jacobs et al., 1992). The carcinogenic potential of the released metal ions and the development of associated neoplasias are still controversial issues. Within this context, the need arises to record cases that will contribute to monitor the potential association between tumor development and placing of a prosthetic structure (Apley, 1989; Brien et al., 1990; Goodfellow, 1992). Features such as ionic valence, particle concentration and size and hypersensitivity have been proposed to explain the potential association between malignant transformation and a metallic implant (Jacobs et al., 1992). In the field of Orthopedics in particular, metallic biomaterials are widely used to manufacture surgical materials such as prostheses for hip replacement or internal fixation devices, and surgeons who deal with traumatic, neoplastic, and degenerative disorders of the skeletal muscle system routinely handle these materials. The potential toxicity of some of the metals most frequently employed in the manufacture of orthopedic implants (titanium, aluminum, vanadium, cobalt, chromium, nickel) has been reported (Elinder & Friberg, 1986; Gitelman, 1989; Jacobs et al., 1991; Jandhyala & Hom, 1983; Langard & Norseth; 1986; Sunderman, 1989; Urban et al., 2000; Williams, 1981). Their carcinogenic potential has been evaluated in animal experimental models (Hueper, 1952; Lewis & Sunderman, 1996; Sinibaldi et al., 1976). The development of tumors at the implant site has been described. Most of the tumors were osteosarcomas or fibrosarcomas associated with stainless steel internal fixation devices (Black, 1988a). However, few reports discuss the potential development of malignant tumors associated to prosthetic structures in humans (Jacobs et al., 1992). Several mechanisms potentially involved in implant-related sarcomatous degeneration have been proposed. However, a direct cause-effect relation between the metal and sarcomatous degeneration in patients has not been demonstrated to date (Black, 1988b; Brown et al., 1987; Case et al., 1996; Goodfellow, 1992). As regards titanium specifically, there are reports of neoplasia in association with dental implants, such as squamous cell carcinoma (Gallego et al., 2008) osteosarcoma (McGuff et al., 2008) and plasmacytoma of the mandible (Poggio, 2007). It is of note that TiO_2 was classified by the International Agency for Cancer Research, as possibly carcinogenic to humans (Group 2B) (Baan et al., 2006).

In this regard, our research group reported a case of sarcomatous degeneration in the vicinity of a stainless steel metallic implant, thus adding to the pool of information that may allow determining more accurately the potential toxicity of metallic implants and the risks associated with their use (Olmedo et al., 2007b).

Regarding the use of titanium and zirconium as implantable materials, Thomsen et al. (1997) found that both titanium and zirconium have a positive effect on tissue-material interaction. A previous experimental study conducted at our laboratory on bone tissue response to an implant, showed greater peri-implant bone thickness and volume in bone surrounding zirconium implants as compared to that around titanium implants (Guglielmotti et al., 1999). Zirconium is chemically related to, and has several properties in common with titanium (Thomsen et al., 1997). According to several researchers (Johansson et al., 1994; Sherepo & Red'ko, 2004; Thomsen et al., 1997; Yuanyuan & Yong, 2007), the elasticity, corrosion resistance, and other physico-mechanical properties of zirconium and its alloys make them a suitable material for biomedical implants. Because zirconium offers superior corrosion resistance over most other alloy systems, better behavior in biological environments can be presumed (Stojilovic, 2005). Nevertheless, it is not widely used as a clinical material at present (Thomsen et al., 1997), since commercial manufacture of implants from zirconium or its alloys seems to be unfeasible due to the high cost of this material (Sherepo & Red'ko, 2004). The potential uses of zirconium-based materials for prosthetics and dental applications should be strongly considered and further investigated in laboratory and clinical settings.

5. Nanotechnology - nanotoxicology

Nanotoxicology is a field of applied sciences involving the control of matter on an atomic or molecular scale, i.e. between 1 and 100 nanometers. Nanotechnology allows creating materials, devices, and systems by controlling matter on a nanometric scale taking advantage of new phenomena and properties (physical, chemical, and biological) that appear at a nanometric scale (Drexler 1986; Mendonçaa et al., 2008).

The aim of applying the principles of nanotechnology to biomaterials (orthopedics and dentistry) is to create materials than can be applied directly to bone tissue, mimicking the natural nanostructure of human tissues by controlling the surface of the implant at a nanometric scale. This would improve the interaction between the implant surface with ions, biomolecules, and cells, favoring the biocompatibility properties of the bioimplant (Mendonçaa et al., 2008). For example, titanium implants with nanostructured coatings, films, and surfaces that seemingly improve the integration of bone tissue with the surface of the implant (osseointegration) and decrease the risk of implant corrosion are currently being developed. Although nanotechnology and its valuable contributions seek to provide answers to the increasing demands of different areas, it is important to understand that these advances may not only bring great advantages but also problems and health risks that must be carefully analyzed and prevented. Thus, nanoparticles may involve deleterious effects to humans or the environment. The fields that study these effects are nanotoxicology (Oberdörster et al., 2005) and nanoecotoxicology (Kahru & Dubourguier, 2009).

Nanoparticles can enter the body by inhalation, ingestion, injection, and/or through the skin (Oberdörster et al., 2005). In addition, they can generate inside the body as occurs when they are released from the surface of metallic implants and biomedical devices, such as coxofemoral prostheses, grids, plates, screws, and distractors used in surgery (Revell, 2006). Little is known about the effect, biodistribution, and final destination of nanometric particles (between 1 and 100 nanometers) inside the body. Given that nanoparticles have a larger surface area per unit of mass compared to microparticles, they may be more bioreactive and potentially more detrimental to human health. Although micro and nanoparticles can be chemically similar, their particular physico-chemical properties such as size, shape, electric charge, concentration, bioactivity and stability, may cause a different biological response. Analyzing the chemistry involved in the release of nanoparticles from metallic surfaces, their size, the quantity that enter the biological milieu, the site where they are transported to, and the immediate and long-term physico-pathological consequences of these particles is a challenge to nanotoxicology and biocompatibility studies (Fig. 7 A-C).

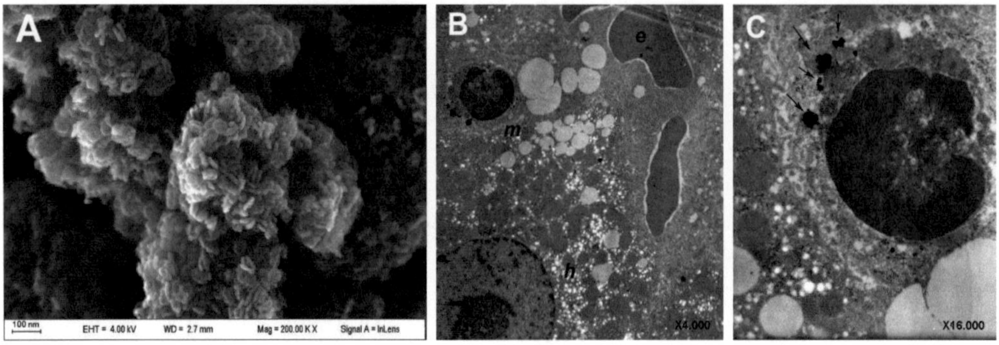

Fig. 7. A) Scanning Electron Microscopy of TiO_2 nanoparticles (10nm). B) Monocyte (*m*) containing nanometric titanium particles (5nm) can be observed in the hepatic sinusoid. (*h*) hepatocyte, (*e*) erythrocyte. X4000. C) Higher magnification allows identifying particles clustered close to the nucleus (→). X16000

6. Conclusions

No metal or alloy is completely inert *in vivo*. Whether noble or passivated, all metals will suffer a slow removal of ions from the surface, largely because of local and temporal variations in microstructure and environment. The potential risk of corrosion and the possible detrimental consequences of corrosion byproducts to tissues are issues of clinical importance.

The biologic effect of corrosion is a public health concern for the community of patients who have a prosthesis (orthopedic and/or dental), since these prostheses remain inside the body over long periods of time.

Evaluation of tissues around metallic devices is important since the presence of ions/particles and their potential local biological effects might affect implant outcome. Corrosion is one of the possible causes of implant failure after initial success. Metal corrosion can affect close contact between the implant and the bone tissue (osseointegration).

The issue of corrosion is not only a local problem since particles resulting from this process could migrate systemically and deposit in target organs. The long term effects of these deposits are yet to be clarified. Mineral elements play a critical role in the physiology and pathology of biological systems. Titanium is a nonessential element; thus, the presence of titanium in the body, titanium biokinetics, and the potential biological effects of titanium are of great interest to researchers.

"In situ" degradation of a metallic implant is an unwanted event since it alters the structural integrity of the implant. Implant manufacturers must attempt to develop methods that reduce the diffusion of metal into the tissues in order to minimize the deleterious effects of corrosion.

We believe further investigation, in particular long-term research, is necessary to advance in the understanding of the factors involved in implant corrosion and establish basic guidelines for their use in clinical implantology. Handling and controlling corrosion of a biomedical implant is essential from a biological, sanitary, metallurgic, economic, and social viewpoint.

Lastly, it is important to highlight that the adverse effects of corrosion described in the present chapter will not invariably occur in all patients with implants since biological response varies among individuals.

7. Acknowledgements

The studies were supported by Grants: PICT 2008-1116 and 1728/OC-AR - PICT 33493 from the National Agency for the Promotion of Science and Technology; UBACyT 20020100200157, UBACyT 20020100100657, UBACyT O-009 and UBACyT O-020 from the University of Buenos Aires; CONICET PIP 6042 and CONICET PIP 11220090100117 from the National Research Council (CONICET) and Roemmers Foundation, Argentina.

8. References

Abraham, J., Grenón, M., Sánchez, H., Pérez, C. & Valentinuzzi, M. (2006). Titanium Based Implants, Metal Release Study in the Oral Environment. LNLS, Activity report. pp. 1-2, ISSN 1518-0204. Available from URL: http://www.lnls.br/ar2006/PDF/909.pdf

Abramson, S., Alexander, H., Best, S., Bokros, J., Brunski, J., Colas, A., Cooper, S., Curtis, J., Haubold, A., Hensch, L., Hergenrother, R., Hoffman, A., Hubbell, J., Jansen, J., King, M., Kohn, J., Lamba, N., Langer, R., Migliaresi, C., More, R., Peppas, N., Ratner, B.,Visser, S., Von Recum, A., Weinberg, S., Yannas, I. (2004). Classes of Materials Used in Medicine, In: *Biomaterials Science. An Introduction to Materials in Medicine,* B.D. Ratner, A.S. Hoffman, F.J. Schoen & J.E. Lemons, (Eds.), pp. 67-233, Elsevier Academic Press, ISBN: 0-12-582463-7, San Diego, California, USA

Adell, R., Lekholm, U., Rockler, B., Bränemark, P. (1981). A 15-year Study of Osseointegrated Implants in the Treatment of the Edentulous Jaw. *International Journal of Oral and Maxillofacial Surgery,* Vol.10, N°6, pp. 387-416, ISSN 0901-5027

Adya, N., Alam, M., Ravindranath, T., Mubeen, A. & Saluja, B. (2005). Corrosion in Titanium Dental Implants: Literature Review. *Journal of Indian Prosthodontic Society,* Vol.5, N°3, (July), pp. 126-131, ISSN 0972-4052

Alfrey, A. (1989). Physiology of Aluminum in Man, In: *Aluminum and Health: A Critical Review,* H.J. Gitelman, (Ed.), pp. 101-124, Marcel Dekker, ISBN 0824780264, New York, USA

Allen, M., Myer, B., Millet, P. & Rushton, N. (1997). The Effects of Particulate Cobalt, Chromium and Cobalt-chromium Alloy on Human Osteoblast-like Cells in Vitro. *The Journal of Bone and Joint Surgery. British volume, Vol.* 79, N°3, (May), pp. 475-482, ISSN 0301-620X

Anderson, J., Cook, G., Costerton, B., Hanson, S., Hensten-Pettersen, A., Jacobsen, N., Johnson, R., Mitchell, R., Pasmore, M., Schoen, F., Shirtliff, M. & Stoodley, P. (2004). Host Reactions to Biomaterials and Their Evaluation, In: *Biomaterials Science. An Introduction to Materials in Medicine,* B.D. Ratner, A.S. Hoffman, F.J. Schoen & J.E. Lemons, (Eds.), pp. 293-354, Elsevier Academic Press, ISBN: 0-12-582463-7, San Diego, California, USA

Apley, A.. (1989). Editorial. Malignancy and Joint Replacement: the Tip of An Iceberg? *The Journal of Bone and joint Surgery. British volume,* Vol.71, N°1, (January), ISSN 0301-620X

Baan, R., Straif, K., Grosse, Y., Secretan, B., El Ghissassi, F. & Cogliano, V. (2006). Carcinogenicity of Carbon Black, Titanium Dioxide and Talc. *The Lancet Oncology,* Vol.7, N°4, (April), pp. 295-296, ISSN 1470-2045

Bessho K, Iizuka T. (1993). Clinical and animal experiments on stress corrosion of titanium miniplates. *Clinical Materials,* Vol.14, N°3, pp. 223-227, ISSN 0267-6605.

Black, J. (1988a). *Orthopedic Biomaterials in Research and Practice,* Churchill Livingstone, ISBN 0443084858, New York, USA

Black J. (1988b). Editorial. Does Corrosion Matter? *The Journal of Bone and joint Surgery. British volume, Vol.*70, N°4; (August), pp. 517-520. , ISSN 0301-620X

Black, J., Sherk, H., Bonini, J., Rostoker, W., Schajowicz, F. & Galante, J. (1990). Metallosis Associated with a Stable Titanium-alloy Femoral Component in Total Hip Replacement. A Case Report. *The Journal of Bone and Joint Surgery. American volume, Vol.*72, N°1, (January), pp. 126-130, ISSN0021-9355

Brånemark, P., Hansson, B., Adell, R., Breine, U., Lindström, J., Hallén, O. & Ohman, A. (1977). Osseointegrated Implants in the Treatment of the Edentulous Jaw. Experience from a 10-year Period. *Scandinavian Journal of Plastic and Reconstructive Surgery. Supplementum,*Vol.16, pp. 1-132, ISSN 0581-9474

Brien, W., Salvati, E., Heley, J., Bansal, M., Ghelman, B. & Betts, F. (1990). Osteogenic Sarcoma Arising in the Area of a Total Hip Replacement: a Case Report. *The Journal of Bone and Joint Surgery. American volume, Vol.*72, N°7, (August), pp. 1097-1099, ISSN 0021-9355

Brodbeck, W., Macewan, M., Colton, E., Meyerson, H. & Anderson, J.M. (2005). Lymphocytes and the foreign body response: lymphocyte enhancement of macrophage adhesion and fusion. *Journal of Biomedical Materials Research. Part A,* Vol.74, N°2, (August), pp. 222-229, ISSN ISSN 1549-3296

Brown, S., Merrit, K., Farnsworth, L. & Crowe, T. (1987). Biological Significance of Metal Ion Release, In: *Quantitative Characterization and Performance of Porous Implants for Hand Tissue Applications,* J.E. Lemon, (Ed.), pp. 163-181, American Society for Testing and Materials, ISBN 0-8031-0965-2, Philadelphia

Bullough, P. (1994). Metallosis. *The Journal of Bone and joint Surgery. British volume,* Vol.76, N°5, (September), pp. 687-688, ISSN 0301-620X

Cabrini, R., Olmedo, D., Tomasi, V. & Guglielmotti, M. (2002). Microincineration for the Detection of Titanium in Tissue Sections. *Journal of Histotechnology, Vol.*25, N°2, pp. 75-78, ISSN 0147-8885

Cabrini, R., Olmedo, D. & Guglielmotti, M. (2003). A Quantitative Method to Evaluate Corrosion Products in Tissues. *Acta Odontológica Latinoamericana,* Vol.16, N°1-2, pp. 27-33, ISSN 0326-481.

Case, C., Langkamer, V., Howell, R., Webb, J., Standen, G., Palmer, M., Kemp, A., Learmonth, I. (1996). Preliminary Observations on Possible Premalignant Changes in Bone Marrow Adjacent to Worn Total Hip Arthroplasty Implants. *Clinical Orthopaedics and Related Research,* 329 Suppl, (August), pp. 269–279, ISSN 0009-921X

Chaturvedi T. (2009). An Overview of the Corrosion Aspect of Dental Implants (titanium and its alloys). (2009). *Indian Society for Dental Research,* Vol.20, N°1, (January-March), pp.91-98, ISSN 0970-9290

Chowdhury, S., Vohra, Y., Lemons, J., Ueno, M. & Ikeda, J. (2007). Accelerating Aging of Zirconia Femoral Head Implants: Change of Surface Structure and Mechanical Properties. *Journal of Biomedical Materials Research. Part B,* Vol.81, N°2, (May), pp. 486-492, ISSN 1552-4973

Daum, S., Anderson, H., Lilis, R., Lorimer, W., Fischbein, S., Miller, A. & Selikoff, I. (1977). Pulmonary Changes Among Titanium Workers. *Proceedings of the Royal Society of Medicine,* Vol.70, N°4, (April), pp. 31-32, ISSN 0035-9157

Diniz-Freitas, M., García-García, A., Crespo-Abelleira, A., Martins-Carneiro, J., Gándara-Rey, J. (2004). Applications of Exfoliative Cytology in the Diagnosis of Oral Cancer. *Medicina Oral, Patología Oral y Cirugía Bucal,* Vol.9, N°4, (August-October), pp. 355-361, ISSN1137-2834

Dowd, J., Schwendeman, L., Macaulay, W., Doyle, J., Shanbhag, A., Wilson, S., Herndon, J., Rubash, H. (1995). Aseptic Loosening in Uncemented Total Hip Arthroplasty in Canine Model. *Clinical Orthopaedics and Related Research,* N°319, (October), pp. 106-121, ISSN 0009-921X

Drexler, K. (1986). *Engines of Creation: The coming Era of Nanotechnology,* Anchor Books, ISBN 0385199732, New York, USA

Duffó, G., Barreiro, M., Olmedo, D., Crosa, M., Guglielmotti, M. & Cabrini, R. (1999). An experimental model to study implant corrosion. *Acta Odontológica Latinoamericana,* Vol.12, N°1, pp. 3-10, ISSN 0326-4815

Effah, E., Bianco, P. & Ducheyne, P. (1995). Crystal Structure of the Surface of Oxide Layer on Titanium and Its Changes Arising from Immersion. *Journal Biomedical Material Research,* Vol.29, N°1, (January), pp.73-80, ISSN 0021-9304.

Elinder, C. & Friberg, L. (1986). Cobalt, In: *Handbook of the Toxicology of Metals,* L. Friberg, G.F.Nordberg & V.B. Vouk, (Eds.), pp. 211-232, Elsevier, ISBN 0444904433, Amsterdam

Engh, C., Moore, K., Vinh, T. & Engh, G. (1997). Titanium Prosthetic Wear Debris in Remote Bone Marrow. A report of Two Cases. *The Journal of Bone and Joint Surgery. American volume,* Vol.79, N°11, (November), 1721-1725, ISSN 0021-9355

Evrard, L., Waroquier, D. & Parent, D. (2010). Allergies to Dental Metals. Titanium: a New Allergen (in French). *Revue médicale de Bruxelles,* Vol.31, N°1, (January-February), pp. 44-49, ISSN0035-3639

Ferguson, A., Laing, P. & Hodge, E. (1960). The Ionization of Metal Implants in Living Tissues. *The Journal of Bone and Joint Surgery. American volume,* Vol.42, (January), pp. 77-90, ISSN 0021-9355

Ferin, J. & Oberdörster, G. (1985). Biological Effects and Toxicity Assessment of Titanium Dioxides: Anatase and Rutile. *American Industrial Hygiene Association Journal,* Vol.46, N°2, (February), pp. 69-72, ISSN 0002-8894

Flatebø, R., Johannessen, A., Grønningsaeter, A., Bøe, O., Gjerdet, N., Grung, B. & Leknes, K. (2006). Host Response to Titanium Dental Implant Placement Evaluated in a Human Oral Model. *Journal of Periodontology,* Vol.77, N°7, (July), pp.1201-1210, ISSN0022-3492

Gallego, L., Junquera, L., Baladrón, J. & Villarreal, P. (2008). Oral Squamous Cell Carcinoma Associated with Symphyseal Dental Implants: an Unusual Case Report. *The Journal of the American Dental Association,* Vol.139, N°8, (August), pp. 1061-1065, ISSN 0002-8177

Garabrant, D., Fine, L., Oliver, C., Bernstein, L. & Peters, J. (1987). Abnormalities of Pulmonary Function and Pleural Disease Among Titanium Metal Production Workers. *Scandinavian Journal of Work, Environment & Health,* Vol.13, N° 1, pp. 47-51, ISSN 0355-3140

Gitelman, H. (1989). *Aluminum and Health: A Critical Review,* Marcel Dekker, ISBN 0824780264, New York, USA

Goodfellow, J. (1992). Editorial. Malignancy and Joint Replacement. *The Journal of Bone and joint Surgery. British volume,* Vol.74, N°5, (September), pp. 645, ISSN 0301-620X

Gottschling, B., Maronpot, R., Hailey, J., Peddada, S., Moomaw, C., Klaunig, J. & Nyska, A. (2001). The Role of Oxidative Stress in Indium Phosphide-induced Lung Carcinogenesis in Rats. *Toxicological Sciences,* Vol.64, N°1, (November), pp. 28-40, ISSN 1096-6080

Guglielmotti, M. & Cabrini R. (1997). Evaluación Biológica de Implantes Dentales Fracasados. Revista de la Asociación Odontológica Argentina, Vol.85, N°4, pp. 313-317, ISNN 0004-4881

Guglielmotti, M., Renou, S. & Cabrini, R. (1999). A Histomorphometric Study of Tissue Interface by Laminar Implant Test in Rats. *The International Journal of Oral & Maxillofacial Implants,* Vol.14, N°4, (July-August), pp. 565-570, ISSN0882-2786

Guindy, J., Schiel, H., Schmidli, F. & Wirz, J. (2004). Corrosion at The Marginal Gap of Implant-Supported Suprastructures and Implant Failure. *The International Journal of Oral & Maxillofacial Implants,* Vol.19, N°6, (November-December), pp. 826-831, ISSN 0882-2786

Hallab, N., Merrit, K. & Jacobs, J. (2001). Metal Sensitivity in Patients with Orthopaedic Implants. Current Concepts Review. *The Journal of Bone and Joint Surgery. American volume,* Vol.83, N°3, (March), pp. 428-436, ISSN 0021-9355

Henrich, V. & Cox, P. (1994). *The Surface Science of Metal Oxides,* Cambridge University Press, ISNN 0521566878, New York, USA

Hueper, W. (1952). Experimental Studies in Metal Carcinogenesis. I. Nickel Cancers in Rats. *Texas Reports on Biology & Medicine, Vol.*10, N°1, pp. 167-186, ISSN 0040-4675

Jacobs, J., Skipor, A., Urban, J. & Galante, R. (1991). Release and Excretion of Metal in Patients Who Have a Total Hip Replacement Component Made of Titanium Base Alloys. *The Journal of Bone and Joint Surgery. American volume;* Vol.73, N°10, pp. 1475-1486, ISSN 0021-9355

Jacobs, J., Rosenbaum, D., Marshall H., Gitelis S. & Black, J. (1992). Early Sarcomatous Degeneration Near a Cementless Hip Replacement. A Case Report and Review. *The Journal of Bone and joint Surgery. British volume,* Vol.74, N°5, (September), pp. 740-744, ISSN 0301-620X

Jacobs, J., Gilbert, J. & Urban, R. (1998). Current Concepts Review - Corrosion of Metal Orthopaedic Implants. *The Journal of Bone and Joint Surgery. American volume,* Vol.80, N°2, (October), pp. 268-282, ISSN 0021-9355

Jandhyala, B. & Hom G. (1983). Minireview. Physiological and Pharmacological Properties of Vanadium. *Life Sciences,* Vol.33, N°14, (October), pp. 1325-1340, ISSN0024-3205

Javed, F., Al-Hezaimi, K., Almas, K. & Romanos, G. (2011). Is Titanium Sensitivity Associated with Allergic Reactions in Patients with Dental Implants? A Systematic Review. *Clinical Implant Dentistry and Related Research,* (March), doi:10.1111/j.1708-8208.2010.00330.x., ISSN1523-0899

Jiranek, W., Machado, M., Jasty, M., Jevsevar, D., Wolfe, H., Goldring, S., Goldberg, M. & Harris, W. (1993). Production of Cytokines Around Loosened Cemented Acetabular Components. Analysis with Immunochemical Techniques and in Situ Hybridization. *The Journal of Bone and Joint Surgery. American volume,* Vol.75, N°6, (June), :863-879, ISSN 0021-9355

Johansson CB, Wennerberg A, Albrektsson T. (1994). Quantitative Comparison of Screw-Shaped Commercially Pure Titanium and Zirconium Implants in Rabbit Tibia. *Journal of Materials Science. Materials in Medicine,Vol.*5, N°6, (June), pp. 340-344, ISSN 0957-4530

Kahru A. & Dubourguier, H. (2009). From Ecotoxicology to Nanoecotoxicology. *Toxicology,* Vol.269, N°2, (March), pp.105-119, ISSN 0300-483X

Kasemo, B. (1983). Biocompatibility of Titanium Implants: Surface Science Aspects. *The Journal of Prosthetic Dentistry,* Vol. 49, N°6, (June), pp. 832-837, ISSN 0022-3913

Kasemo, B. & Lausmaa, J. (1998). Biomaterial and Implant Surfaces: A Surface Science Approach. *The International Journal of Oral & Maxillofacial Implants,* Vol.3, N°4, (Winter), pp.247-259, ISSN0882-2786

Katou, F., Andohn, N., Motegui, K. & Nagara, H. (1996). Immuno-Inflammatory Responses in the Tissue Adjacent to Titanium Miniplates Used in the Treatment of Mandibular Fractures. *Journal of Cranio-maxillo-facial Surgery,* Vol.24, N°3, (June), pp.155-162, ISSN1010-5182

Kawanishi, S., Oikawa, S., Inoue, S. & Nishino, K. (2002). Distinct Mechanisms of Oxidative DNA Damage Induced by Carcinogenic Nickel Subsulfide and Nickel Oxides. *Environmental Health Perspectives,* Vol.110, N°5, (October), pp. 789-791. 0091-6765

Koppang HS, Roushan A, Srafilzadeh A, Stølen SØ & Koppang R. (2007). Foreign Body Gingival Lesions: Distribution, Morphology, Identification by X-ray Energy Dispersive Analysis and Possible Origin of Foreign Material. *Journal of Rral Pathology and Medicine,* Vol.36, N°3, (March), 161-172, ISSN *0904-2512*

Kumazawa, R., Watari, F., Takashi, N., Tanimura, Y., Uo M. & Totsuka, Y. (2002). Effects of Ti Ions and Particles on Neutrophil Function and Morphology. *Biomaterials,* Vol.23, N°17, pp.3757-3764, ISSN 0142-9612

Laing, P. Compatibility Of Biomaterials. (1973). *The Orthopedic Clinics of North America,* Vol.4, N°2, (April), pp. 249-273, ISSN 0030-5898

Lalor, P., Revell, P., Gray, A., Wright, S., Railton, G. & Freeman, M. (1991). Sensitivity to Titanium. A Cause of Implant Failure? *The Journal of Bone and joint Surgery. British volume, Vol.*73, N°1, (January), pp. 25-28, ISSN 0301-620X

Lang, N., Wilson, T. & Corbet, E. (2000). Biological Complications with Dental Implants: Their Prevention, Diagnosis and Treatment. *Clinical Oral Implants Research,* Vol.11, Suppl 1, pp. 146-155, ISSN 0905-7161

Langard, S. & Norseth, T. (1986). Chromium, In: *Handbook of the Toxicology of Metals,* L. Friberg, G.F.Nordberg & V.B. Vouk, (Eds.), pp. 185-210, Elsevier, ISBN 0444904433, Amsterdam

Langkamer, V., Case C., Heap P., Taylor A., Collins C., Pearse M., & Solomon L. (1992). Systemic Distribution of Wear Debris After Hip Replacement. A Cause for Concern? *The Journal of Bone and joint Surgery. British volume,* Vol.74, N°6, (November), pp. 831-839, ISSN 0301-620X

Lee K, Henry NW, Trochimowicz HJ, Reinhardt CF. (1986). Pulmonary Response to Impaired Lung Clearance in Rats Following Excessive Tio_2 Dust Deposition. *Environmental Research,* Vol.41, N°1, (October), pp. 144-167, ISSN 0013-9351

Lee, J., Salvati E., Betts F., DiCarlo E., Doty S. & Bullough P. (1992). Size of Metallic and Polyethylene Debris Particles in Failed Cemented Total Hip Replacements. *The Journal of Bone and joint Surgery. British volume,* Vol.74, N°3, (May), pp. 380-384, ISSN 0301-620X

Leggett, R. (1989). The Behavior and Chemical Toxicity of U in the Kidney: A Reassessment. *Health Physics,* Vol.57, N°3, (September); pp.365-383, ISSN0017-9078

Lewis, C., & Sunderman F. (1996). Metal Carcinogenesis in Total Joint Arthroplasty. Animal Models. *Clinical Orthopaedics and Related Research,* N°329, (August), pp. 264-268, ISSN 0009-921X

Long, M. & Rack, H. (1998). Titanium Alloys in Total Joint Replacement: A Materials Science Perspective. *Biomaterials,*Vol. 19, N°18, (September), pp. 1621-1639, ISSN 0142-9612

Lothar, P., Weili, L.,& Horst, H. (1992). Effect of Fluoride Prophylactic Agents on Titanium Surfaces. *The International Journal of Oral & Maxillofacial Implants,*Vol.7, N°3, (Fall), pp. 390-394, ISSN0882-2786

Lu J, Descamps M, Dejou J, Koubi G, Hardouin P, Lemaitre J, Proust JP. (2002). The Biodegradation Mechanism of Calcium Phosphate Biomaterials in Bone. *Journal of Biomedical Materials Research. Part A,* Vol.63, N°4, pp. 408-412, ISSN1549-3296

Lucas, L. & Lemons, J. (1992). Biodegradation of Restorative Metallic Systems. *Advances in Dental Research,* Vol.6, (September), pp.32-37, ISSN 0895-9374Manicone, P., Rossi Iommetti, P. & Raffaelli, L. (2007). An Overview of Zirconia Ceramics: Basic Properties and Clinical Applications. *Journal of Dentistry, Vol.*35, N°11, (November), pp. 819-826, 0300-5712

Matthew, I., Frame, J., Browne, R. & Millar, B. (1996). In Vivo Surface Analysis of Titanium and Stainless Steel Miniplates and Screws. *International Journal of Oral and Maxillofacial Surgery, Vol.*25, N°6, (December), pp. 463-468, ISSN0901-5027

Matthew, I. & Frame, J. (1998). Allergic Responses to Titanium. *Journal of Oral and Maxillofacial Surgery, Vol.*56, N°2, (December), pp.1466-1467, ISSN 0278-2391

Maziere, C., Floret, S., Santus, R., Morliere, P., Marcheux, V. & Maziere, J. (2003). Impairment of The EGF Signaling Pathway by the Oxidative Stress Generated with UVA. *Free Radical Biology & Medicine,*Vol.34, N°6, (March), pp. 629-636. ISSN 0891-5849

McGuff, H., Heim-Hall, J., Holsinger, F., Jones, A., O'Dell, D. & Hafemeister, A. (2008). Maxillary Osteosarcoma Associated with A Dental Implant: Report of a Case and Review of the Literature Regarding Implant-Related Sarcomas. *The Journal of the American Dental Association,* Vol.139, N°8, pp. 1052-1059, ISSN 0002-8177

Meachim, G. & Williams, D. Changes In Nonosseous Tissue Adjacent to Titanium Implants. (1973). *Journal of Biomedical Materials Research, V*ol.7, N°6, (November), pp. 555-572, ISSN 0021-9304 (Print) 1097-4636

Meffert, R., Langer, B. & Fritz, M. (1992). Dental Implants: A Review. *Journal of Periodontology,* Vol.63, N°11, (November), pp. 859-870, ISSN0022-3492

Mendonçaa, G., Mendonçaa, D., Aragãoa, F. & Cooperb, L. (2008). Advancing Dental Implant Surface Technology - From Micron- to Nanotopography. *Biomaterials,* Vol.29, N°28, (October), 3822-3835, ISSN 0142-9612

Merritt, K., Brown, S. & Sharkey, N. (1984). The Binding of Metal Salts and Corrosion Products to Cells and Proteins In Vitro. *Journal of Biomedical Materials Research. Part A,* Vol.18, N°9, (November-December), pp. 1005-1015.

Moberg LE, Nordenram Å, Kjellman O. (1989). Metal Release from Plates Used in Jaw Fracture Treatment. A Pilot Study. *International Journal of Oral and Maxillofacial Surgery,* Vol.18, N°5, (October), pp. 311-314, ISSN 0901-5027

Moran, C., Mullick, F., Ishak, K., Johnson, F. & Hummer, W. (1991). Identification of Titanium in Human Tissues: Probable Role In Pathologic Processes. *Human Pathology,* Vol.22, N°5, (May), pp. 450-454, ISSN 0046-8177

Nalli, G., Verdú, S., Paparella, M., Olmedo, D & Cabrini, R. (2009). Exfoliative Cytology and Titanium Dental Implants. A Preliminary Study. *Journal of Dental Research* (in press)

Newman, M. & Flemming, T. (1988). Periodontal Considerations of Implants and Implant Associated Microbiota. *Journal of Dental Education, Vol.*52, N°12, (December), pp. 737-744, ISSN0022-0337

Nikolopoulou, F. Saliva and Dental Implants. (2006). *Implant Dentistry,* Vol.15, N°4, (December), pp. 372-376. ISSN 1056 6163

Oberdörster, G., Oberdörster, E. & Oberdörster J. (2005). Nanotoxicology: An Emerging Discipline Evolving from Studies of Ultrafine Particles. *Environmental Health Perspectives,*Vol.113, N°7, (July), pp. 823-839, ISSN 0091-6765

Olmedo, D., Guglielmotti, M., Cabrini, R. (2002). An Experimental Study of the Dissemination of Titanium and Zirconium in The Body. *Journal of Materials Science. Materials in Medicine, Vol.*13, N°8, (August) 793-796, ISSN 0957-4530

Olmedo, D., Fernández, M., Guglielmotti, M. & Cabrini, R. (2003a). Macrophages Related to Dental Implant Failure. *Implant Dentistry;* Vol.12; N°1, pp.75-80, ISSN 1056-6163

Olmedo, D., Tasat, D., Guglielmotti, M. & Cabrini, R. (2003b). Titanium Transport Through the Blood Stream. An Experimental Study In Rats. *Journal of Materials Science. Materials in Medicine, Vol.*14, N°12, pp. 1099-1103, ISSN 0957-4530

Olmedo, D., Tasat, D., Guglielmotti, M. & Cabrini, R. (2005). Effect of Titanium Dioxide on The Oxidative Metabolism of Alveolar Macrophages: An Experimental Study In Rats. *Journal of Biomedical Materials Research. Part A,*Vol.73, N°2, (May), pp. 142-149, ISSN1549-3296

Olmedo, D., Paparella, M., Brandizzi, D., Spielberg, M. & Cabrini. (2007a). R. Response of Oral Mucosa Associated to Titanium Implants. *Journal of Dental Research 86 (Spec Iss B):* 0083. Available from URLhttp://iadr.confex.com/iadr/arg07/preliminaryprogram/abstract_114433.htm

Olmedo, D., Michanié, E., Olvi, L., Santini-Araujo, E. & Cabrini, R. (2007b). Malignant Fibrous Histiocytoma Associated to Coxofemoral Arthrodesis. *Tumori,* Vol.93, N°5, (September-October), pp. 504-507, ISSN 0300-8916

Olmedo, D., Tasat, D., Evelson, P., Guglielmotti, M. & Cabrini, R. (2008a). Biological Response of Tissues with Macrophagic Activity to Titanium Dioxide. *Journal of*

*Biomedical Materials Research. Part A, Vol.*84, N°4, (March), pp. 1087-1093, ISSN1549-3296

Olmedo, D., Tasat, D., Evelson, P., Guglielmotti, M. & Cabrini, R. (2008b). Biodistribution of Titanium Dioxide from Biologic Compartments. *Journal of Materials Science. Materials in Medicine,* Vol.19, N°9, (September), pp. 3049-3056, ISSN 0957-4530

Olmedo, D., Duffó, G., Cabrini, R. & Guglielmotti, M. (2008c). Local Effect of Titanium Implant Corrosion: An Experimental Study in Rats. *International Journal of Oral and Maxillofacial Surgery,* Vol.37, N°11, (November), pp.1032-1038, ISSN 0901-5027

Olmedo, D., Tasat, D., Duffó, G., Guglielmotti, M. & Cabrini, R. (2009). The Issue of Corrosion in Dental Implants: A Review. *Acta Odontológica Latinoamericana,* Vol.22, N°1, pp. 3-9, ISSN 0326-4815

Olmedo, D., Paparella, M., Brandizzi, D. & Cabrini, R. (2010). Reactive Lesions of Peri-Implant Mucosa Associated with Titanium Dental Implants: A Report Of 2 Cases. *International Journal of Oral and Maxillofacial Surgery,* Vol.39, N°5, (May), pp. 503-507, ISSN 0901-5027

Olmedo, D., Tasat, D., Evelson, P., Rebagliatti, R., Guglielmotti, M. & Cabrini, R. (2011). In Vivo Comparative Biokinetics and Biocompatibility of Titanium and Zirconium Microparticles. *Journal of Biomedical Materials Research. Part A,* Vol.98, N°4, (September), pp. 604-613, ISSN 1549-3296

Poggio, C. (2007). Plasmacytoma of the Mandible Associated with a Dental Implant Failure: A Clinical Report. *Clinical Oral Implants Research,* Vol.18, N°4, (August), pp. 540-543, ISSN 0905-7161

Ratner, B., Hoffman, A., Schoen, F. & Lemons, J. (2004). Biomaterials Science: A Multidisciplinary Endeavor, In: *Biomaterials Science. An Introduction to Materials in Medicine,* B.D. Ratner, A.S. Hoffman, F.J. Schoen & J.E. Lemons, (Eds.), pp. 1-9, Elsevier Academic Press, ISBN: 0-12-582463-7, San Diego, California, USA

Revell, P. (2006). The Biological Effects of Nanoparticles. *Nanotechnology Perceptions,* Vol.2, N°3, pp. 283-298, ISSN 1660-6795

Rosenberg, A., Grätz, K. & Sailer, H. (1993). Should Titanium Miniplates Be Removed After Bone Healing is Complete? *International Journal of Oral and Maxillofacial Surgery,* Vol.22, N°3, (June), pp.185-188, ISSN 0901-5027

Schliephake, H., Lehmann, H., Kunz, U., Schmelzeisen & R. (1993). Ultrastructural Findings in Soft Tissues Adjacent to Titanium Plates Used in Jaw Fracture Treatment. *International Journal of Oral and Maxillofacial Surgery, Vol.*22, N°1, (February), pp. 20-25, ISSN0901-5027

Sherepo, K. & Red'ko, I. (2004). Use of Zirconium-Based and Zirconium-Coated Implants in Traumatology and Orthopedics. *Biomedical Engineering,* Vol.38, N°2, (March), 77-79, ISSN 1475-925X

Sicilia, A., Cuesta, S., Coma, G., Arregui, I., Guisasola, C., Ruiz, E. & Maestro, A. (2008). Titanium Allergy in Dental Implant Patients: A Clinical Study on 1500 Consecutive Patients. *Clinical Oral Implants Research,* Vol.19,N°8, (August), pp. 823-835, ISSN 0905-7161

Siirilä, H. & Könönen, M. (1991). The Effect Of Oral Topical Fluorides On The Surface of Commercially Pure Titanium. *Journal of Oral & Maxillofacial Implants,* Vol.6, N°1, (Spring), pp. 50-54, ISSN0882-2786

Sinibaldi, K., Rosen, H., Liu, S. & DeAngelis, M. (1976). Tumors Associated with Metallic Implants in Animals. *Clinical Orthopaedics and Related Research,* N°118, (July-August), 257-266., ISSN 0009-921X

Solheim, E., Sudmann, B., Bang, G. & Sudmann, E. (2000). Biocompatibility and Effect on Osteogenesis of Poly(Ortho Ester) Compared to Poly(DL-Lactic Acid). *Journal of Biomedical Materials Research. Part A,* (February), Vol.49, N°2, pp. 257-263, ISSN 1549-3296

Sollazzo, V., Palmieri, A., Pezzetti, F., Bignozzi, C., Argazzi, R., Massari, L., Brunelli, G &, Carinci, F. (2007). Genetic Effect of Zirconium Oxide Coating on Osteoblast-Like Cells. *Journal of Biomedical Materials Research. Part B, Vol.*84, N°2, (February), pp.550-558, ISSN 1552-4973

Stojilovic, N., Bender, E. & Ramsier, R. (2005). Surface Chemistry Of Zirconium. *Progress in Surface Science,* Vol. 78, N°3, pp. 101-184, ISSN 0079-6816

Sul, Y., Johansson, C., Jeong, K., Roser, K., Wennerberg, A. & Albrektsson, T. (2001). Oxidized Implants and Their Influence on Their Bone Response. *Journal of Materials Science. Materials in Medicine,* Vol.12, N°10-12, (October-December), pp. 1025-1031, ISSN 0957-4530

Sunderman, F. (1989). A Pilgrimage Into The Archives of Nickel Toxicology. *Annals of Clinical and Laboratory Science,* Vol.19, N°1, (January-February), pp. 1-16, ISSN 0091-7370

Tagger Green, N., Machtei, E., Horwitz, J. & Peled, M. (2002). Fracture of Dental Implants: Literature Review And Report Of A Case. *Implant Dentistry;* Vol. 11, N°2, pp. 137-143, ISSN 1056-6163

Takebe J, Champagne CM, Offenbacher S, Ishibashi K, Cooper LF. (2003). Titanium Surface Topography Alters Cell Shape and Modulates Bone Morphogenetic Protein 2 Expression In The J774A.1 Macrophage Cell Line. *Journal of Biomedical Materials Research. Part A,* Vol.64, N°2, (February), pp. 207-216, ISSN 1549-3296

Textor, M., Sittig, C., Frauchiger, V., Tosatti, S. & Brunette, D. (2001). Properties and Biological Significance of Natural Oxide Films on Titanium and Its Alloys, In: *Titanium in Medicine,* D.M. Brunette, P. Tengvall, M. Textor, P. Thomsen, (Eds.), pp. 171-230, Springer-Verlag, ISBN 3540669361, Berlin, Germany

Thomsen, P., Larsson, C., Ericson, L., Sennerby, L., Lausmaa, J. & Kasemo, B. (1997). Structure of The Interface Between Rabbit Cortical Bone and Implants Of Gold, Zirconium And Titanium. *Journal of Materials Science. Materials in Medicine,* Vol.8, N°11, pp. 653-665, ISSN 0957-4530

Torgersen, S., Gjedet, N., Erichsen, E. & Bang, G. (1995). Metal Particles and Tissue Changes Adjacent to Miniplates. A Retrieval Study. *Acta odontologica Scandinavica,* Vol.53, N°2, (April), pp. 65-71, ISSN0001-6357

Urban, R., Jacobs, J., Gilbert, J. & Galante, J. (1994). Migration of Corrosion Products from Modular Hip Prostheses. Particle Microanalysis and Histopathological Findings. *The Journal of Bone and Joint Surgery. American volume,* Vol.76, N°9, (September), pp. 1345-1359, ISSN 0021-9355

Urban, R., Jacobs, J., Tomlinson, M., Gavrilovic, J., Black, J. & Peoc'h, M. (2000). Dissemination of Wear Particles to The Liver, Spleen, and Abdominal Lymph Nodes of Patients with Hip or Knee Replacement. *The Journal of Bone and Joint Surgery. American volume, Vol.*82, N°4, (April), pp. 457-476, ISSN 0021-9355

Williams, D. (1981). Biological Effects of Titanium, In: *Systemic Aspects of Biocompatibility,* D.F. Williams, (Ed.), pp. 169-177, CRC Press, ISBN 0849366216, Boca Ratón, Florida, USA

Williams, D. (1982). Corrosion of Orthopaedic Implants, In: *Biocompatibility of Orthopaedic Implants,* D.F. Williams, (Ed.), pp. 197-229, CRC Press, ISBN 0849366135, Boca Ratón, Florida, USA

Williams, D. (1987). Definitions in Biomaterials, *Proceedings of a Consensus Conference of the European Society for Biomaterials,* ISBN 0444428585, Chester, England, March 3-5, 1986

Woodman, J., Black, J. & Jimenez, S. (1984a). Isolation of Serum Protein Organometallic Corrosion Products from 316LSS And HS-21 in Vitro and in Vivo. *The Journal of Bone and Joint Surgery. American volume,* Vol.18, N°1, (January), pp. 99-114, ISSN 0021-9355

Woodman, J., Jacobs, J., Galante, J. & Urban, R. (1984b). Metal Ions Release from Titanium-Based Prosthetic Segmental Replacements of Long Bones in Baboons: A Long-Term Study. *Journal of Orthopaedic Research,* Vol.1, N°4, pp. 421-430, ISSN *0736-0266*

Xia, Z. & Triffitt J. (2006). A Review on Macrophage Responses to Biomaterials. *Biomedical Materials,* Vol.1, N°1; pp. 1-9, (March), ISSN 1748-6041

Yang, J. & Merrit, K. (1994). Detection of Antibodies Against Corrosion Products in Patients After Co-Cr Total Joint Replacements. *Journal of Biomedical Materials Research. Part A,* Vol.28, N°11, (November), 1249-1258., ISSN1549-3296

Young-Kyun, K., Hwan-Ho, Y. & Seung-Cheul, L. (1997). Tissue Response to Titanium Plates: A Transmitted Electron Microscopic Study. *Journal of Oral and Maxillofacial Surgery,* Vol.55, N°4, (April), pp. 322-326, ISSN 0278-2391

Yuanyuan, Y. & Yong, H. (2007). Structure and Bioactivity of Micro-arc Oxidized Zirconia Films. *Surface and Coatings Technology, Vol.*201, N°9, (February), pp. 5692-5695, ISSN.

Zaffe, D., Bertoldi, C., Consolo & U. (2003). Element Release from Titanium Devices Used in Oral and Maxillofacial Surgery. *Biomaterials, Vol.*24, N°6, (March), pp. 1093-1099, ISSN 0142-9612

Zirconium. *Mineral Information Institute.* 2009. Available from URL: http://www.mii.org/Minerals/photozircon.html

6

Importance of Etch Film Formation During AC Controlled Pitting of Aluminium

Maria Tzedaki[1], Iris De Graeve[1], Bernhard Kernig[2], Jochen Hasenclever[2] and Herman Terryn[1]
[1]Vrije Universiteit Brussel,
[2]HYDRO Aluminium Bonn
[1]Belgium
[2]Germany

1. Introduction

Over the last decades much research has been dedicated and reported in literature on the Alternating Current electrograining (A.C. electrograining) mechanism and the parameters that change the final pitting morphology of electrograined aluminium. The reason is the increasing industrial demand every year (currently estimated to be 800,000,000 m^2/year) for high quality litho-printing and super capacitors for energy storage. Both applications relay on the production of a controlled roughened surface on aluminium foil which can be achieved through A.C. electrograining. In this case pitting is used as a surface treatment for the production of a larger surface area with uniform pits.

In the past, research focused on different conditions of the A.C. electrograining of aluminium in relation to the final pitting morphology. The charge density, the wave shape of the applied current, the electrolyte concentration, the frequency or the temperature are factors that were studied extensively. The industrial application of the A.C. electrograining process makes it very important to understand and correlate all the parameters which affect the graining. The mechanism which initiates or affects the final morphology needs to be understood. In this chapter we will present an overview on the influence of the different A.C. electrograining conditions on the final pitting morphology, with the focus mainly on the smut film formation. The latter has proven to play an important role during the A.C. electrograining but little correlation was shown so far between the smut film formation and the pitting morphology. We will present in this article the importance of the smut film formation mechanism during the A.C. electrograining of aluminium and the H_2 evolution, taking place during the cathodic half cycle. Conditions, such as the applied potential, the electrolyte concentration, the frequency and the charge density are important to be presented since they play a crucial role in the creation of the final pitting morphology of aluminium.

Much effort has recently been put towards understanding the H_2 evolution; therefore it is important to include in this chapter reported differences observed in gas retention within the etch smut film for different electrochemical conditions and relate these differences to the final smut morphology.

The aim of this article is to gather information available on the etch smut morphology and the smut formation mechanism for a better understanding of the influence of the etch film on the final pitting morphology.

2. A.C. electrograining mechanism

A.C. electrolytic graining or electrograining is a process used for the creation of a uniformly pitted morphology on aluminium surfaces. Electrograining can increase the specific area of a smooth aluminium foil by up to a factor of 200 [Ono & Habazaki, 2011, Dyer & Alwitt 1981]. The process is industrially applied in lithographic printing plates, where a larger surface area is necessary in order to enhance the water retentive properties of the aluminium and to improve the adhesion of the photosensitive coating which is deposited after electrograining and anodising of the aluminium substrate. Also in the preparation of aluminium capacitor foils an electrochemical etching of aluminium is required. This enlarges the surface area of the electrolytic capacitor electrodes permitting a reduction both in cost and in size of the capacitor [Jackson, 1975].

The aluminium substrate is electrograined in an electrolytic cell using a suitable electrolyte, usually a strong acid like hydrochloric acid (HCl) or nitric acid (HNO_3). Additives such as citric acid, nitric acid, boric acid etc. are used to improve the uniformity of the pits [Terryn, 1987]. The required morphology is created by applying a sinusoidal alternating current (Figure 1) or voltage which determines the initiation, propagation and repassivation of key pits during each period of the AC cycle, mostly with a frequency of 50 or 60 Hz.

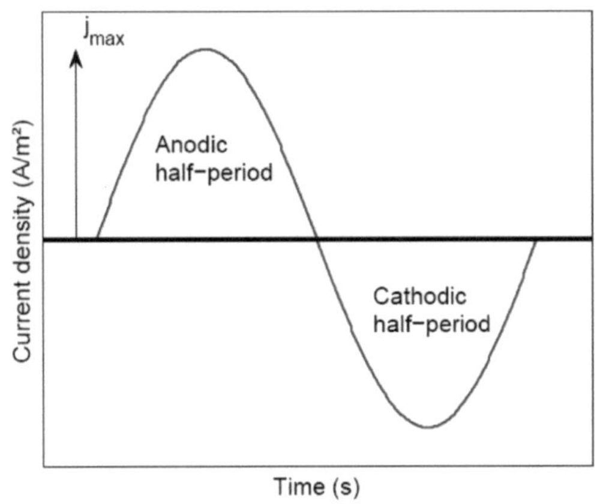

Fig. 1. Applied alternative AC current waveform during one period

During the anodic half cycle oxidation of aluminium to Al^{+3} ions occurs; dissolved aluminium is removed to the bulk of the solution and cubic pits are formed [Dyer & Alwitt, 1981]. The reactions taking place during the positive polarisation are:

$$Al \rightarrow Al^{3+} + 3e^- \quad (1)$$

$$2H_2O \rightarrow O_2 + 4H^+ + 4e^- \quad (2)$$

In the cathodic half period, due to hydrogen gas evolution (reaction 3), the pH can rise above 5. The previously formed Al^{3+} ions which were not yet removed from the surface, precipitate as aluminium hydroxide $Al(OH)_3$ (reaction 4). Aluminium hydroxide is insoluble in water and precipitates as a white gel [Vargel et al, (2004), Dyer & Alwitt, 1981]. The latter together with aluminium particles, water and ions from the electrolyte create an amorphous etch film, called smut layer, formed in a colloidal phase which masks the pits and causes an increase in the potential. [Terryn, 1987, 1988, 1991a, 1991b, Laevers, 1992, 1995]

$$2H^+ + 2e^- \rightarrow H_2 \quad (3)$$

$$Al^{3+} + 3OH^- \rightarrow Al(OH)_3 \quad (4)$$

A further increase of the pH above 9 will result in oxidation of aluminium to AlO^-_2 ions (reaction 5); these will eventually precipitate as aluminium hydroxide when the pH of the layer close to the surface decreases rapidly [Pourbaix, 1974, Terryn, 1987, Laevers, 1995, Buytaert, 2006, Wilson et al, 2008, Hülser, 1955]. The decrease of the pH can be due to protons (H^+) coming from the bulk electrolyte and/or oxygen evolution during the abovementioned reactions. [Laevers, 1995, Buytaert, 2006, Pourbaix, 1974, Terryn, 1987,]

$$Al + 4OH^- \rightarrow AlO^-_2 + 2H_2O + 3e^- \quad (5)$$

$$AlO^-_2 + 2H_2O \rightarrow Al(OH)^-_4 \quad (6)$$

$$Al(OH)^-_4 \rightarrow Al(OH)_3 + OH^- \quad (7)$$

$$4OH^- \rightarrow O_2 + 2H_2O + 4e^- \quad (8)$$

The cathodic half period involves the re-passivation of the substrate resulting in a re-distribution of attack during the subsequent anodic half-period, which is necessary for the creation of a uniform pitting morphology [Dowell, 1979, Terryn, 1987, Buytaert, 2006]. A schematic representation of the electrograining process can be seen in Figure 2.

Depending on the electrolyte, the pit initiation and propagation can differ and the pitting morphology can vary between a uniform, non-uniform and etch-like morphology. The formation of pits can also vary between crystallographic, shallow and worm-like pits as illustrated in Figure 3.

The resulting pit shape depends on whether the process is activation or diffusion controlled. During the activation controlled process, metal atoms are removed from various crystal planes resulting in the formation of crystallographic pits [Smialowska, 1986]. This condition occurs during the first stages of pit growth where only small changes in Cl^- concentration occur within the pit [Smialowska, 1986]. When chloride ions undergo chemisorptions on the crystallographic facets on the bottom of the pits the potential increases and solution concentrates within the pits. Here diffusion is the predominant phenomenon resulting in the formation of hemispherical pits [Smialowska, 1986].

The properties of the passive film influence the initiation of the pit but they do not contribute significantly in the pit growth. For stable pit growth a high concentration of Cl^- and a low pH inside the pit eliminate the repassivation at the initial stage. After longer times the pit growth is a diffusion controlled process [Smialowska, 1999].

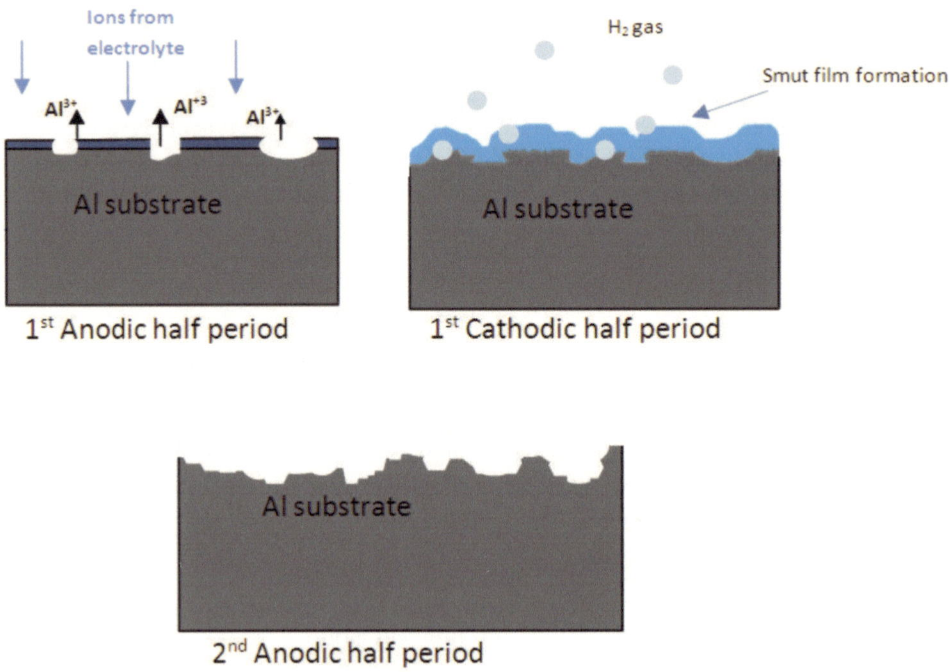

Fig. 2. A.C. electrograining mechanism of aluminium

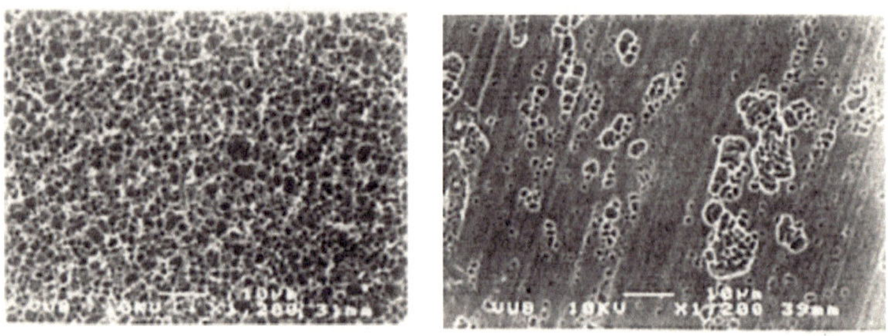

Fig. 3. Shallow (left) and worm-like (right) pitting of Al 99.99 in HNO_3, 50Hz [Laevers, 1995]

Koroleva showed that by controlling the extent of anodic or cathodic polarisation and by considering the local interfacial solution conditions, tailored and optimised pitting morphologies are obtained. This can be understood by examining the cathodic polarisation which promotes the pit initiation and eliminate pit growth; a high population of fine hemispherical pits is observed after electrograining with cathodic bias [Koroleva, 2005].

How the electrograining conditions can influence the uniformity of the pits and the final surface morphology will be discussed in the following paragraphs.

2.1 Influence of the electrolyte on the electrograining process

For a uniform pitting morphology on an aluminium substrate, a suitable electrolyte must be chosen. A strong inorganic acid such as hydrochloric or nitric acid has been mainly used and studied extensively the past years. Each one can give a uniform pitting morphology at a suitable temperature and concentration and with the appropriate applied current density. The solutions used for graining are diluted and there should be a balance between the aggressiveness of the electrolyte, necessary to initiate the pitting, and the low anion concentration, necessary for the re-passivation to occur [Dowell, 1986]. HCl is suitable at a range of 0.5 to 10% v/v whereas less aggressive acids need a higher concentration [Dowell, 1986].

HCl gives a characteristic crystallographic pitting morphology whereas HNO_3 results in a shallower pitting structure [Laevers et al, 1993, Terryn 1987, Bridel et al, 1983]. The density and the number of pits are proportional to the electrolyte aggressiveness and the substrate conditions, whereas the pit size is related to the local anodic charge passed [Laevers et al, 1993, Terryn, 1988, Dyer & Alwitt, 1981, Dowell, 1979, 1986]. The use of Hydrofluoric Acid leads to the formation of surfaces that are made of significantly smaller, shallower and more uniform pits with a much higher pit density as it is illustrated in Figure 4 [Wilson, 2006]. The presence of hydrofluoric acid in solution leads to both a rapid dissolution of the second phase particles and a breakdown of the protective oxide film with pit initiation as a result of the presence of highly aggressive F^- ions in the solution [Wilson, 2006].

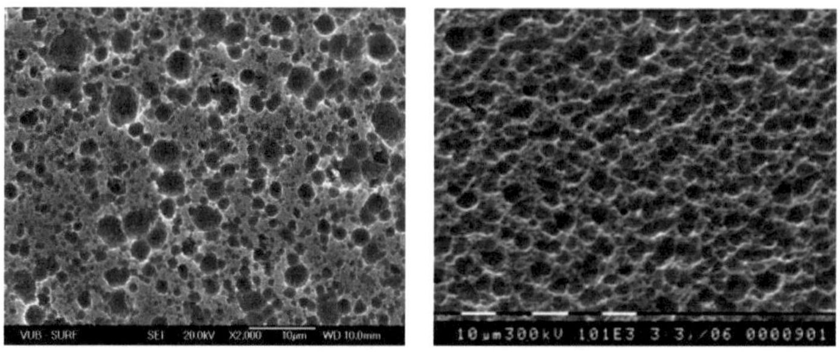

Fig. 4. Pitting morphology of AA 1050 after A.C. electrograining in 12.5g HCl (left) without HF and with addition of 1.25g/l Hydrofluoric Acid (right)

Several additives are also important especially in industrial processes, such as acids and aluminium salts e.g. aluminium chloride or aluminium nitrate.

The use of additives such as citric acid, disodium phenyl phospate dihydrate (DPPD), acetic acid etc. can improve the uniformity of the pits [Terryn, 1987]. Acetic acid can catalyse the hydrogen evolution reaction. Acetate ions are involved in inhibiting the reactions at the surface either by surface deposition or as a competitor ion to chloride. The use of the DPPD as an additive decreases the dissolution reaction of aluminium, and polarisation data has shown a shift in the pitting potential to higher values. The increased surface inhibition by the addition of DPPD is caused by the formation of a protective layer on the aluminium.

An interesting study on the influence of the additives during the electrograining was done by Wilson et al, (2008) who evaluated through NMR studies the differences induced by additives in a hydrochloric acid electrograining solution regarding the smut and the pitting morphology. The addition of DPPD or citric acid was proven to deteriorate or enhance respectively the pit uniformity. Further, the addition of citric acid causes the formation of a more porous smut layer and a final surface morphology with larger pits and fewer plateau areas in comparison with the morphology produced in HCl electrolyte.

The addition of H_2SO_4 decreases the pit cluster size and increases pit-on-pit growth. Sulfate ions that are absorbed on the surface can impede the nucleation and the pit-on-pit propagation [Ono & Habazaki 2009]. A suitable sulfate concentration in the etchant can lead to a deep and homogenously etched layer with high surface area [Ono & Habazaki 2009]. It was shown previously that sulfate ions could shift the pitting potential in the positive direction [Tomcsanyi et al 1989, Kim et al 1999, Lee et al 2000, Ono & Habazaki 2009]. This was later interpreted as the elimination of the competitive adsorption of sulfate ions by chloride ions [Ono & Habazaki 2009]. The rise of the pitting potential indicates the impediment of pit initiation of pure aluminium by sulfate ions [Kim et al, 1999, Ono & Habazaki 2009]. It was also observed that an excess of H_2SO_4 in the electrolyte promotes the formation of an etched layer with uniform thickness, since the pit formation is suppressed on the aluminium surface [Ono & Habazaki 2009].

Lin and Li, (2006) on the other hand had a different observation regarding the pitting formation in the presence of sulfate ions. They observed that the formation of strings of cubic pits was suppressed by the sulfate ions and the pit formation of the aluminium surface was enhanced.

2.2 Potential evolution during the electrograining process

Dimogerontakis, (2007) and Laevers, (1995) studied the changes in anodic and cathodic potentials during the electrograining, proposing a mechanism for the electrograining in HCl. It was observed by Dimogerontakis that within the first second of the electrograining three subregions are observed (Figure 5). Within the first few ms both the anodic and cathodic potential increases strongly due to the dissolution of the second phase particles that are distributed at the surface and the possible build up of dissolution products. In the second region the anodic potential decreases to a minimum value, which is correlated to the degradation of the initial protective oxide film present on the surface and to the initiation of the crystallographic pitting of the aluminium surface. A third region is observed where both anodic and cathodic potential increases mainly due to the deposition of etch product during the negative half period, which begins at the end of the second region. Moreover the cathodic reaction has a greater hindrance from the smut than the aluminium dissolution.

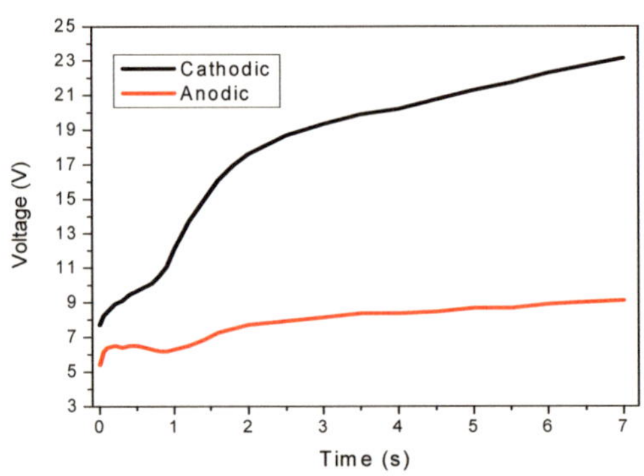

Fig. 5. Anodic and Cathodic peaks of the potential vs. time during the A.C. electrograining in HCl

The amount of charge passing through the surface during the anodic half cycle is given by the following equation [Laevers et al, 1992, 1993]

$$\frac{Qperiod}{2} = \int_0^{\pi} J_{max} \cdot \sin\omega t \cdot dt = -\frac{J_{max}}{\omega}[\cos\omega t]_0^{\pi}$$

and the local anodic charge [Laevers et al, 1992, 1993] is

$$q_{pit} = \frac{\frac{q_{period}}{2}}{PitNumber}$$

where J_{max} (A/m^2) is the maximum current density and $\omega=2\pi f$, with f(Hz) the electrograining frequency.

From the previous two equations it can be concluded that by decreasing the frequency of the graining the period of time during which the anodic polarisation takes place increases, while an increase in the current density promotes the oxidation of aluminium [Laevers et al, 1993].

Laevers et al, (1992, 1993, 1998) mention that the response potential (cell voltage including the ohmic drop over the electrolyte) during the electrograining increases as a function of increasing current density, but show that in HCl the response potential is independent of the current density imposed. This leads to the conclusion that in HCl the electrograining of aluminium was controlled by aluminium oxidation and hydrogen gas evolution, which is a mass transport phenomenon, and thus by the transport conditions of reactants and products at the electrode surface. The presence of the smut layer has an influence on the mass transport phenomena affecting the electrical properties of the electrode/solution interface.

Lee et al (2000) showed that the exposed metal surface at the bottom of the micro-pits that are formed during the cathodic polarization acts as a preferential site for Cl^- ions attack during the following anodic half cycle. Thus with a sufficient cathodic polarization the formation of micro pits can take place, promoting the corrosion of pure aluminium [Lee et al 2000].

Ono and Habazaki, (2010) proved that there is no influence of the current waveform (direct, anodic or square) on the amount of dissolved aluminium, but it changes the etch morphology from angular pits (direct current) to large hollow pits at low density (anodic current) and uniform etch films (for a square wave).

2.3 Influence of the substrate on the electrograining process

Most of the commercial aluminium alloys have a relatively high concentration of magnesium, iron, silicon and manganese, therefore it is important to show the influence that those elements have on the electrochemical properties of aluminium.

It was observed that local, coarse pit development is due to the presence of intermetallic particles [Terryn, 1987]. The intermetallics can act as cathodic sites and therefore anodic dissolution can take place at the interface of the intermetallic with the aluminium matrix [Terryn, 1987, 1988]. In this way the locally concentrated current reduces the amount of the effective current available for electrograining and thus for pit initiation or propagation [Terryn, 1987, 1988].

Dowell, (1986) mentioned that intermetallic particles (such as $FeAl_3$, αAl-Fe-Si and in less activity βAl-Fe-Si) act as sites for pit initiation. He also showed that grain boundaries act as preferential sites for the pit development whereas alloy elements such as copper, iron and zinc give more plateau areas in the final pitting morphology. Mg, Si and Mn did not appear to have a significant influence on the pitting morphology. Even though many pits are initiated at the sites of intermetallic particles there were many particles (around 50%) that were observed as inactive in the pit generation. The mechanism proposed for the initiation at intermetallic particles was that intermetallics do not form a protective oxide layer, therefore even if they are electrochemically less negative than the aluminium they dissolve faster revealing the underlying aluminium surface, which can then act as initiation point.

Laevers, (1996) presented the influence of manganese on the electrograining of aluminum. A more uniform final pitting morphology was observed when the amount of manganese present in solid solution was decreased, but this could only be noticed in the more advanced stages of the graining process.

On the influence of iron and silicon Laevers, (1995) showed that when iron is present in solid solution, the average pit is shallower, whereas when silicon is present a non uniform morphology is observed with merged hemispherical pits, worm like pits and unattacked surface.

Zn and Mg promote a quick initiation of pits due to the more effective current distribution, which leads to a convoluted surface with larger grains and eliminated plateau areas [Sanchez et al, 2010].

Little information is available regarding the grain boundaries and the grain orientation in relation to specific preferential sites for the pitting initiation. It was observed that there were only 30% more pits at boundaries than expected from a random distribution [Dowell, 1986]. Cubic pits on the other hand are characteristic of the attack on (100) crystallographic planes because impurities segregate to cellular boundaries in [100] direction [Martinez-Caicedo et al 2002, Koroleva et al, 2005]. The segregates enhance cathodic activity, increasing the response potential [Martinez-Caicedo et al, 2002]. It was also observed that the subgrain boundaries could be the pit nucleation sites at the onset of electrograining [Marshall et al, 1995].

Two effects of microcrystallization on the pitting behaviour of pure aluminium are reported: (1) the rate of pit initiation is accelerated; (2) the pit growth process is impeded. This leads to the enhancement of pitting resistance for microcrystallized aluminium [Meng et al, 2009]

During the industrial rolling of aluminium sheet, rolled-in oxides are formed, creating a disturbed subsurface layer [Buytaert, 2006]. This alters the surface morphology of the aluminium substrate and thus influences the electrograining in HCl or HNO_3 solutions. It was observed that a large amount of ungrained rolled-in oxides remains even after graining at high charge densities [Rodriguez, 2011]. The interface between rolled-in oxides and aluminium matrix can act as a weak point for the pit initiation, therefore an etching was proposed as a necessary pre-graining step [Rodriguez, 2011].

3. Smut film formation and morphology

It is known that during electrograining an etch film is formed masking the aluminium surface, mainly composed of hydrated aluminium (Figure 6) [Thompson & Wood, 1978, Dowell 1986, Terryn, 1987, Marshall, 1995 Laevers, 1996, Amor & Ball, 1998]. The film formation mechanism, morphology and weight differ upon different graining conditions therefore a thorough investigation of several parameters is necessary.

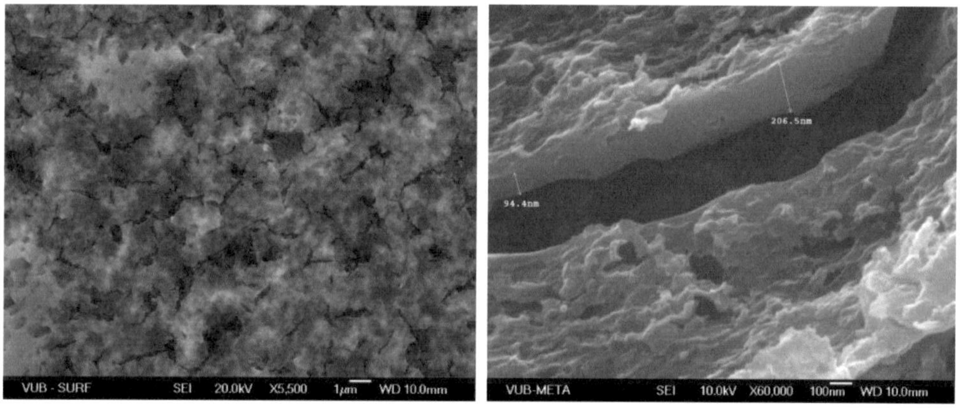

Fig. 6. Smut film morphology of aluminium sheet electrograined in HCl at 50Hz top view(left) and bended & cracked sample(right) [Raes, 2009]

Dowell, (1979) showed that the cell voltage undergoes a proportional increase to the weight of smut film formed during graining in HCl. This indicates that the smut film causes an increase in the resistance to current flow.

As it was observed by Dyer and Alwitt, (1981) cubes propagate from new different point gaps inside each cube due to the weak points in the protective etch film. Whereas during direct curent etching of alumium under the same conditions no smut is formed, and cubes are formed by subsequent propagation outward of the first cube. The presence of the smut changes the pit formation mechanism and prevents the outward propagation. This indicates that the smut layer plays an important role in the final pitting morphology. The thickness and the morphology of the smut are yet to be examined in order to fully identify their role.

Laevers, (1995) observed that by imposing the same conditions during electrograining in HCl and in HNO_3 electrolytes, the amount of etch product is the same but the morphology differs. The smut formed in HCl is a non uniform mud-like layer whereas in HNO_3 smut is formed at the rim of the regular hemispherical pits [Laevers, 1993]. He also mentioned that if the amount of Al^{3+} ions, precipitated as aluminium hydroxyde is low, hemispherical pits with walls decorated with a high population density of cubic shaped key pits will be formed. If a high amount of hydroxyde is precipitated, an etch like morphology will be obtained. This indicates that the development of the surface morphology is influenced by the mechanism of development of the etch layer.

In the presence of HF a thin and highly porous etch film with numerous superfine pits on the surface is observed [Wilson, 2006]. Only a very thin, non-uniform layer of smut can be deposited in the presence of HF as a result of both the highly aggressive nature of the F^- ions present and the production of highly soluble corrosion products compared to those produced in HCl [Wilson, 2006]. The activation of the surface by the fluoride ions and the lack of formation of a compact etch film allows the whole of the aluminium surface to be available to undergo graining reactions [Wilson, 2006]. Unlike the case of the hydrochloric acid where a build up of smut can lead to some blocking of the surface, especially in the regions proximal to the early pits that are formed [Wilson, 2006]. This means that as the electrograining process progresses the area susceptible to graining is reduced leading to continued reaction at pre-existing pits, which in turn leads to the observation of larger and deeper pit morphologies [Wilson, 2006].

The etch film itself is amorphous [Lin et al, 2001, Laevers, 1995] but tends to crystallize when irradiated by 200 keV electrons. Smut is formed during the cathodic half cycle of the electrograining by a dissolution/precipitation mechanism caused by the production of H_2 resulting in high local pH values [Thompson, 1978].

In recent studies it was observed that the addition of additives on the electrolyte can influence the smut morphology. The addition of acetic acid is shifting the pH in lower values thus producing a smut with very small pores [Wilson, 2008]. The porous smut layer increases the amount of aluminium dissolution or pit initiation resulting in a final surface morphology with less plateau areas [Wilson, 2008]. On the other hand it was shown that the addition of DPPD shifts the interfacial pH to higher values, reducing the dissolution rate of aluminium; therefore a pitting morphology with few but large pits surrounded by large plateau areas is observed [Wilson, 2008].

Dimogerontakis et al, (2006) noticed that by increasing the current density the amount of produced smut increases as a result of the high concentration of aluminium inside the pits during the anodic half period and of the local increase of the pH during the cathodic half cycle. This resulted in a decrease of the population density of the pits and to an enhanced pit growth. Thus at high current we can obtain less but bigger pits.

Lin and Chiu, (2005) showed that the variation of the duration time of the anodic, t_a to cathodic, t_c half cycle changes the thickness of the smut. Smut is decreased by increasing the t_a/t_c ratio while the average size of the individual hemispherical pits increases. This indicates that during the anodic half cycle the pits grow through the dissolution of aluminium while the etch film is formed during the cathodic half cycle when local pH increases due to hydrogen gas evolution. Moreoever they noticed that the etch film in some areas was pushed away from the surface by certain forces exerced on the interface. It is believed that the pressure accumulated within the hydrogen bubbles or produced when H_2 gas leaves the smut surface gives space for new pit initiation in the aluminium substrate during the anodic half cycle. When the etch film reaches a certain thickness, the conduction resistance increases, the further pit growth will be prevented and consequently new pits will be initiated elsewhere.

3.1 Hydrogen gas evolution

The presence of hydrogen gas through reaction 2.3 can cause the increase of the local pH, which eventually causes the precipitation of aluminium hydroxide and the formation of the smut layer, masking the pitting morphology. In this paragraph we will try to understand the mechanism of the hydrogen gas evolution while it is released and leaves the smut and identify whether there is a relation between the H_2 evolution, the morphology of the smut and the pitting morphology.

The reduction of H^+ during the cathodic half period was observed to be a mass transport controlled phenomenon [Laevers, 1995]. This indicates that the amount of precipitated aluminium hydroxide will be influenced by the time needed for sufficient H^+ to be gathered at the surface for the pH to rise above 5. The amount of the precipitated aluminium hydroxide is determined by the surface area, which is taking part in the active dissolution of aluminium [Laevers, 1995].

Hydrogen gas evolution during the cathodic half period may cause the transition of the pit initiation from flaws of the film to sites by removal or cracking up of the protective oxide film. By increasing the treatment time or the charge density, the growth rate is not the same for all pits.

Few recent publications can be found regarding the hydrogen gas evolution during the cathodic half cycle of the electrograining of aluminium.

Tomasoni et al, (2010) showed that by using a rotating aluminium disk as working electrode, the hydrogen bubble size and the potential during the electrograining in HCl are increasing at the same rate and show the same transient in time. As the pit diameter increases, the bubble departure time from the electrode decreases whereas during region III (as indicated by Dimogerontakis and Terryn, (2007)) it was noticed that the smut layer is filled, at decreasing rate, by gas while growing: the hydrogen flux from the electrode to the bulk solution is thus increasing in time and the bubble break-off diameter increases as well.

Small Angle X-ray Scattering (SAXS) is proposed to be able to characterize $Al(OH)_3$ gels [Bale & Schmidt, 1959, Christensen et al, 1982, Sinko et al, 1999]. The aluminium hydroxide particles inside precipitated gels have been considered as oblate spheroids (platelets) or polydisperse spheres that are made from individual (solid) aggregated $Al(OH)_3$ molecules [Bale & Schmidt, 1959, Bottero, 1980, 1982, Christensen et al, 1982, Bradley, 1993, Sinko et al, 1999]. Those aggregate particles can agglomerate inside the gel as seen from SAXS measurements [Bale & Schmidt, 1959, Sinko et al, 1999].

It is known that smut is an amorphous $Al(OH)_3$ gel, with a dry thickness less than 1 μm [Thomspon & Wood, 1978, Terryn et al, 1991a, 1991b, Lin et al, 2001]. In recent studies using small angle x-ray scattering it was shown that hydrogen is retained in the smut, playing an important role in the process [Hammons et al, 2010]: scattering data can be presented as a log to log plot of the intensity, I, versus the scattering vector, q, which is related to the angle of each intensity measurement. The size, shape and structure of particles can be determined from SAXS data. During the electrograining there is significant solution bubbling which increases the transmission compared to subsequent frames where there is lower transmission due to lower amount of gas bubbles in the beam path. The first SAXS data showed that there is a significant volume fraction of gas within the gelatinous smut layer in HCl [Hammons et al, 2010]. When DPPD was used as an additive in the electrolyte, less hydrogen fraction was detected in the smut so it was concluded that hydrogen can permeate through the smut in a more uniform manner without damaging it to a significant extend and therefore the dissolution occurs in a more homogeneous way [Hammons et al, 2010].

The reason why gas is retained in the smut is yet to be elucidated but it is assumed to be related to the gas release during or after the electrograining [Hammons, 2011]. Gas release after electrograining and gas diffusion can be estimated through SAXS measurements, but gas permeation during the electrograining is difficult to study [Hammons, 2011].

Gas release after electrograining is characterized by some parameters correlated to the gas diffusivity of the smut and to the total gas release [Hammons, 2011]. These parameters are considered to be functions of how well the smut can accommodate gas release in the absence of a pressure drop [Hammons, 2011].

A mechanism for the gas release was proposed by Hammons, (2011) through data collected from scattering images using in situ Small Angle X-ray Scattering (SAXS). It has been suggested that pores or water passageways that can facilitate gas permeation during electrograining close after the electrograining, and as such the gas is trapped inside the smut under certain conditions [Hammons, 2011] as illustrated in Figure 7. However, at a higher frequency and with the incorporation of additives, pores do not collapse and form a tortuous porous network that causes an increase in the pressure of the hydrogen gas near the interface [Hammons, 2011].

By observing the pitting morphologies after desmutting of the samples it was seen that the interfacial gas pressure can influence the final pitting morphology and give a more etch like morphology, as was observed beneath smut with little porosity, whereas at higher smut porosity a more uniform morphology is observed [Hammons, 2011].

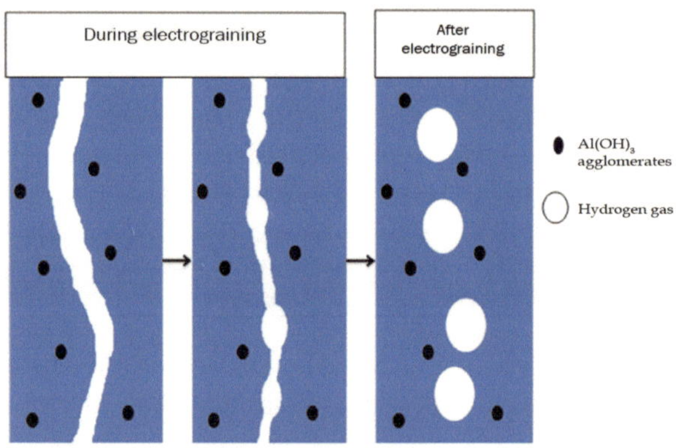

Fig. 7. Mechanism for smut permeation during electrograining where a passageway opens around the $Al(OH)_3$ agglomerates and after electrograining the pores collapse leaving behind retained gas (white) [Hammons, 2011]

The use of additives and the gas release behavior was also examined with SAXS. Smut formed with citric acid has similar gas diffusion properties of that with HCl [Hammons 2011]. Based on the presence of large pores, high gas diffusion and low gas retention, smut was formed with DPPD [Hammons 2011]. These large pores are correlated with a surface morphology that contained very deep pits [Hammons 2011]. This work is yet to be published.

A difference in the smut porosity was observed with different additives as reported previously. The additives increase the smut porosity, compared to the reference sample that was grained only in HCl. The addition of citric acid resulted in a smooth smut morphology, similar to HCl, but with more porosity. Although a gas release mechanism similar to that of a HCl electrograined sample was experimentally observed in the SAXS data, an increase in the measured potential indicated that some open pore permeation was possible and could explain the difference in the final pit morphology, compared to the reference HCl electrograined sample [Hammons, 2010, 2011].

Electrochemical impedance spectroscopy was also used to describe the pitting formation mechanism. Brett, (1990) proposed a multistep dissolution mechanism by considering that surface aluminium atoms are oxidised stepwise until a soluble aluminate film is formed. It was concluded that migration takes place and therefore in a transient solution the soluble products are removed easily while increased mass transport supplies larger quantities to the electrode surface [Brett, 1990]. In an upcoming publication we will show that the Multisine Electrochemical Impedance Spectroscopy can give more valuable information regarding the presence of hydrogen trapped inside the smut. In addition to the known effect of additives such as DPPD, which are known to give a different smut thickness and morphology, it is assumed that those additives also influence the gas permeation mechanism. By using impedance data the movement of the gas inside the smut layer can be modelled.

3.2 Pitting morphology

HCl is known to give a uniform crystalographic cubic shape pitting morphology whereas HNO_3 gives more shalow like pits [Terryn, 1987, Laevers, 1995] as it is illustrated in Figure 8.

It was observed by Terryn, (1987) that by increasing the frequency from 1 to 100 Hz we can observe a change from morphologies with locally attacked sites and discrete pits (1Hz) to more uniform attacked areas with increased pitting density. A sufficient current can give uniform pitted surfaces but when a current density of 300 A_{rms} dm^{-2} is applied it can give rise to a less uniform pitting morphology [Laevers, 1995].

Thompson and Wood, (1978) showed that in HCl, pit formation occurs at gaps in the air formed alumina film on the aluminium substrate. The gaps can be either caused mechanically due to physical features on the metal or can be associated with compositional features [Thompson & Wood, 1978, Lindseth 1999, Afseth, 1999]. Upon graining in HCl a large population density of pits (more than 10^{13} m^{-2}) of sizes around 160 nm is formed in the first cycles of electrograining and remains constant, but the smaller pits grow during the A.C. electrograining and merge with other pits leaving less plateau areas [Thompson & Wood, 1978].

By comparing the shape of the key pits building up the surface we could conclude that the anodic metal dissolution during graining in hydrochloric acid occurs in preferred directions, while during the dissolution of aluminium in nitric acid it occurs isotropically (building up from a flat walled hemispherical pit) [Laevers 1995]. The anodic and cathodic process seems to occur on fresh surfaces at new positions for each cycle and the cube propagation starting at different points in the cubic phase appears to happen in a random way [Dyer & Alwitt, 1981].

Pitting is preferentially following the rolling lines as observed from the first electrograining steps [Dowell, 1979, 1986]. Grain boundaries act to some extend as preferential sites for the initiation of pits [Dowell, 1979, 1986].

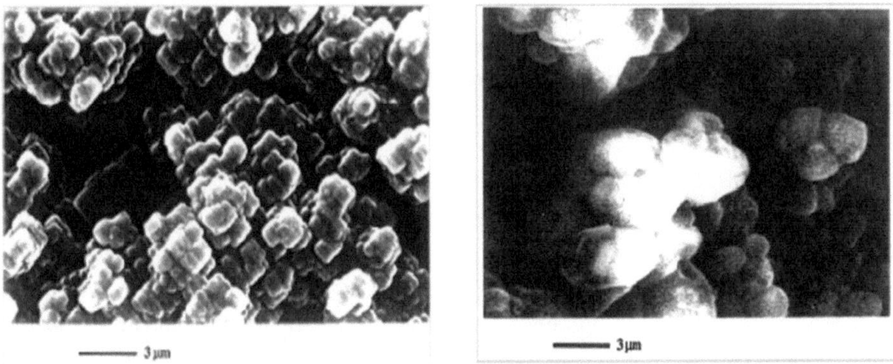

Fig. 8. Oxide replicas of pits formed during electrograining in HCl (left) with cubic pit building up and HNO_3 (right) with flat walled hemispherical pit building up[Laevers, 1995]

The effective size of the building elements increases by increasing current density and decreasing graining frequency [Terryn et al, 1988, 1991a, 1991b]. The walls of hemispherical pits become smooth as the size of cubic pits decreases with increasing frequency [Terryn et al, 1988, 1991a, 1991b].

Intermetalics present in the substrate influence the graining mechanism. Many pits are initiated at sites of intermetalics, especially in the presensce of Mg, Fe, Zn and Cu, but not all particles have the same active role during the electrograining [Dowell, 1986].

For 1 wt% of manganese alloyed aluminum plate with most of the manganese (0.9 wt%) present in solid solution, a lateral and unidirectional growth of hemispherical pits occurs, creating worm-like pits [Laevers et al, 1996]. It was assumed that the worm-like pits observed are also influenced by other alloying elements present in solid solution, such as iron, silicon and titanium for AA1050 aluminum [Laevers et al, 1996].

In general it can be concluded that HCl and HNO_3 can give different pit shapes, but in both cases during the anodic half period a large number of pits is created simultaneously and with time growth or new pits are initiated and at a certain moment some pits will merge together creating bigger pits (Figure 9). The final surface morphology can vary between uniform morphologies with homogeneously distributed pits to areas with plateaus or a non uniform distribution of pits.

3.2.1 Pit size

Electrograining in HCl gives at the first cycles large hemispherical pits in different sizes which become broader and grow till they merge [Thompson & Wood, 1978, Terryn et al 1988, 1991a, 1991b, Dimogerontakis et al, 2006, 2007].

After electrograining in HCl and examining the surface at different times Lin et al (2001) reported the different evolution during the graining of the fine, hemispherical and worm like pits. Within the first 60 seconds they showed that the small dotted-like pits grow and coalescence with the fine pits to form more hemispherical pits. While the process continues the worm pits merge with the growing hemispherical pits and disappear. After around 120 seconds most of the aluminium surface is dotted with large hemispherical pits with a uniform distribution which can lead upon further graining into the growth and merger with other hemispherical pits, forming pit sizes of 5μm. In the same study cross-sectional images showed that the surface of the worm like pits is dotted with fine hemispherical pits whereas the hemispherical pits are observed with flat walls. The fine pits have a smaller depth than the hemispherical ones, which indicates that the growth of the hemispherical pits prevails and propagates into the aluminium substrate.

An increase in the current density can decrease the pit population density, but increases the mean value of the pit size [Dimogerontakis et al, 2006]. With an increased current a higher concentration of aluminium ions inside each pit is obtained during the anodic half cycle making the pore of the pit bigger while during the cathodic half cycle a larger precipitation of smut takes place due to the higher local pH increase [Dimogerontakis et al, 2006].

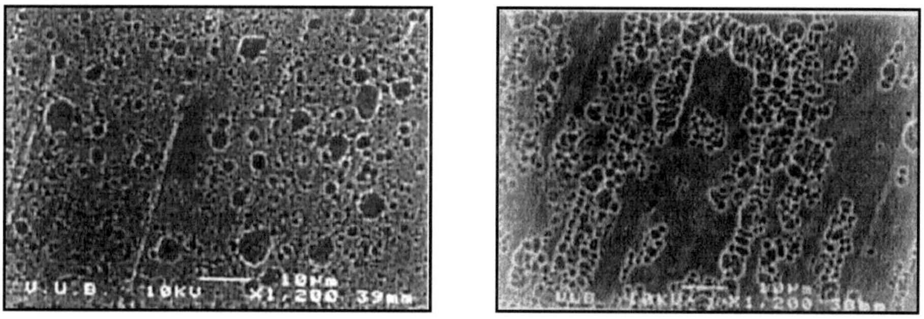

Fig. 9. Pitting morphology after electrograining with HCl (left) and HNO_3 (right) nder the same electrograining conditions [Laevers, 1995]

3.2.2 Correlation with smut

A very important parameter for the development of graining morphologies is the number of sites where key pits are initiated, mainly determined by the surface condition of the aluminium substrate and the propagation rate of the sites, depending on the features of the applied alternating current [Laevers, 1995].

It was observed that the weight of the smut layer increases linearly with the consumed electric charge during graining in hydrochloric acid [Dimogerontakis et al, 2006]. By applying a high electric charge the formation of an increased smut layer will cause the repassivation of the metal resulting in the increase of the cathodic potential [Terryn 1987, Dimogerontakis et al, 2006]. A more intensive hindrance during the cathodic reaction could be due to the fact that the H_2 evolution was observed to be mass transport controlled [Laevers, 1995]. In this way the formation of a thick hydrated smut layer can hinder the reaction of the hydrogen ions [Dimogerontakis et al, 2004, 2006]. An increase of the current density results in a smut layer formation with increasing size of hemispherical pits [Dimogerontakis et al, 2006]. On the other hand, a less uniform graining morphology was developed by increasing the current density [Dimogerontakis et al, 2006]. The lower the amount of formed smut the more uniform graining morphology is obtained with small hemispherical pits, a high pit population density and a more uniform size distribution.

4. Conclusion

Electrograining is an important process for the lithographic printing plate industry and for aluminium capacitor foil production. The purpose of electrograining is to create a higher surface area with a convoluted surface of pits with narrow size distribution, shallow depths and no ungrained or plateaus areas.

During electrograining the aluminium surface is continuously undergoing a process of pitting and repassivation. In the anodic polarisation which occurs during the first half cycle of the electrograining period, pit formation enhanced by chloride or nitrate ions is observed while during the cathodic half cycle an increase of the pH due to H_2 production occurs. The latter leads to the precipitation of $Al(OH)_3$ on the surface creating a layer called smut. The smut is a gel that contains 5% $Al(OH)_3$, 5% Al and 90% water.

It is clearly seen that the understanding of the electrograining mechanism was and still remains a subject for extended research. All the different parameters analyzed give different information regarding the reactions taking place, the dissolution of aluminium, the formation of the pits, the smut deposition and the final morphology.

In this paper we presented studies mainly focusing on electrograining in HCl or HNO_3 electrolytes. HCl is one of the most common used electrolytes for the electrograining of aluminium. It gives a characteristic crystallographic pitting morphology whereas a shallower pitting structure can be achieved when electrograining in HNO3 electrolyte occurs. In general both electrolytes give similar overall surface morphologies under certain electrograining conditions, but the morphology of the individual pit is still different. HF is known to give a high density of small and uniform pits. At high electrolyte concentrations and low current densities etch-like morphologies are obtained whereas with increasing current density and decreasing concentration, uniform and hemispherical pits are observed, with sizes depending on the frequency. At very low concentrations and very high current densities, a non-uniform pitting morphology is obtained because hemispherical pits grow laterally until they intersect. On the other hand with an increasing electrolyte flow across the electrodes, the surface morphology will tend towards an etch-like morphology, with no hemispherical pits.

The addition of acids can change the surface morphology. Similar morphologies to HCl were seen with the addition of acetic acid giving larger pits with finer features whereas the addition of citric acid results in shallower hemispherical pits. Significant difference was observed with the addition of DPPD with very deep pits and large plateau areas.

The initial substrate surface morphology can influence the final pitting morphology as it was observed in many studies. Pit initiation and nucleation were observed to occur at metallurgical, mechanical or compositional flaws in the aluminium surface. Intermetallic particles and especially β-phase particles can act as cathodic sides promoting a local coarse pit development. The presence of manganese or silicon in solid solution gives less uniform pitting morphologies whereas when iron is present shallower pits are created. Zinc and magnesium due to a fast pitting initiation give surfaces with larger pits and eliminate plateau areas.

A thin and porous etch film leads to a homogenous grained surface due to a homogenous distribution of current density. A thick and compact film containing few defects causes the formation of large pits and extended plateaus as a result of current concentration at the defects. With high current we can obtain less but bigger pits due to the increased amount of produced smut as a result of the high concentration of aluminium inside the pits during the anodic half period.

An important parameter for the creation of the smut layer is the presence of hydrogen at the surface of the aluminium. The amount of precipitated aluminium hydroxide will be influenced by the time needed for a sufficient amount of H^+ to be gathered at the surface for the pH to rise above 5. The mechanism through which hydrogen gas escapes from the smut and is released in the bulk of the solution was described through SAXS data. It was proposed that when gas is trapped inside the smut a tortuous porous network is observed giving an increased pressure on the interface. This results in a more etch-like pitting morphology whereas when there is sufficient smut porosity and gas release a more uniform pitting morphology can be obtained.

A more thorough investigation regarding the smut film formation mechanism, the structure and the morphology of the smut in correlation to the hydrogen gas evolution during the cathodic polarisation on the electrograining of aluminium is necessary in order to be able to describe and correlate the influence of the hydrogen upon the final pitting morphology. Electrochemical Impedance data can give an inside observation of the retained gas; this is ongoing research yet to be published.

5. Acknowledgment

The authors would like to thank Josh Hammons and Yves Van Ingelgem for the support and the help they provided for the preparation of the article.

6. References

Afseth, A. (1999). Metallurgical control of filiform corrosion of aluminium alloys, NTNU, Ph.D. Thesis, Trondheim, Norway

Akitt, J., W. (1989). Progress in Nuclear Magnetic Resonance Spectroscopy, Volume 21, pp.1-149.

Amor, M., P. & Ball, J. (1998). The mechanism of electrograining aluminium sheet in Nitric/Boric acid electrolyte, Corrosion Science, volume 40, pp. 2155-2172

Bale, H., D. & Schmidt, P., W. (1959). Small angle X-ray mattering from aluminium hydroxide gels, Physical Chemistry, Volume 31, pp.1612-1618.

Bottero, J., Y., Cases, J. M., Fiessinger, F. & Poirier, J., E. (1980). Studies of hydrolysed aluminum chloride solutions, Chemical Physics, Volume 84, pp.2933-2939.

Bottero, J., Y., Tchoubar, D., Cases, J., M. & Flessinger, F. (1982). , Investigation of the hydrolysis of aqueous solutions of aluminum chloride. 2. Nature and structure by small-angle x-ray scattering, Physical chemistry, Volume 86, pp. 3667-3673.

Bradley, S., M., Kydd, R., A., & Howe, R., F. (1993). The structure of Al-gels formed through base hydrolysis of Al^{3+} aqueous solutions, Colloid Interface Science.Volume 159, pp. 405-412.

Brett, C., M., A. (1990). The application of lectrochemical impedance techniques to aluminium corrosion in acidic chloride solution, Applied Electrochemistry, Volume 20, pp. 1000-1003

Bridel, F., Grynszpan, R., Bourelier, F., Bavay, J., C. & Vu Quang, k. (1983). Passivity of metals and semiconductors, Elsevier, Amsterdam, p. 753

Buytaert, G., (2006). Study of (Sub)surface on rolled commercially pure Aluminium alloys, VUB, Ph.D. Thesis, Brussels, Belgium

Christensen, A., N., Lehmann, M., S. & Wright, A. (1982). Acta Chemica Scandinavica Series a-Physical and Inorganic Chemistry, Volume 36, pp. 779-781.

Dimogerontakis, Th., Campestrini, P. & Terryn, H. (2004). The influence of the current density on the AC-graining morphology, Electrochemical Society Proceedings, Volume 19

Dimogerontakis, Th., Terryn, H. & Campestrini, P. (2006). Repassivation of aluminium during AC-graining process by aluminium hydrohide formation, *9th International Symposium: Passivation of Metals and Semiconductors, and Properties of Thin Oxide Layers*, Paris, France, June 27- July 1, 2005

Dimogerontakis, Th., & Terryn, H. (2007). Interpretation of anodic and cathodic potential variations during AC-graining of aluminium in hydrochloric acid, Corrosion Science, Volume 49, pp. 3428-3441

Dowell, A., J. (1979). Alternating current etching of aluminium lithographic sheet, Transactions of the Institute of Metal Finishing, Volume 57), pp. 138–144

Dowell, A., J. (1986). The influence of metallurgical factures in A.C. etching of lithoplates, Transactions of the Institute of Metal Finishing, pp. 85-90.

Dyer, C., K. & Alwitt, R., S. (1981). Surface changes during A.C. etching of aluminium, Electrochemical Society, 128, 300-305.

Hammons, J., Raymenta, T., Vandendael, I., Blajiev, O., Hubin, A., Davenport, A., J., Raes, M. & Terryn, H. (2010). A method to detect retained gas during AC electrograining using in-situ small angle X-ray scattering, Electrochemistry Communications, Volume 12, pp. 717-719

Hammons, J. (2011). A study of precipitated films formed during electrochemically driven dissolution process, University of Birmingham, Ph.D. Thesis, Birmingham, United Kingdom,

Hebert, K. & Alkire, R., C. (1988), Growth and passivation of aluminum etch tunnels, Electrochemical Society, Volume 135, pp. 2146-2157

Hülser, P., Krüger, U., A. & Beck, F. (1955). The cathodic corrosion of aluminium during the electrodeposition of paint: Electrochemical measurements, Corrosion Science, Volume 38, pp. 47-57

Jackson, N., F. (1975). The tunnel etching of aluminium, Electrocomponent Science and Technology, Volume 2, pp 33-44

Johansson, G. (1960). On the Crystal Structure of Some Basic Aluminium Salts, Acta Chemica Scandinavica, Volume 14, pp. 771-773.

Kim, S., S., Lee, W., J., Pyun, S., I. & Kim, D., R. (1999). Effects of applied potential and solution temperature on the pitting corrosion of pure aluminium in sulphate ion containing chloride solution, Metals and Materials. Volume 5, pp. 583–588.

Koroleva, E., V., Thompson, G., E., Skeldon, P., Hollrigl, G., Smith, G. & Lockwood, S. (2005). Tailored AA1050 alloy surfaces by electrograining, Electrochimica Acta, Volume 50, pp. 2091-2106

Laevers, P., Terryn & H., Vereecken, J. (1992). Comparison of the A.C. electrograining of aluminium in hydrochloric and nitric acid, Transactions of the Institute of Metal Finishing, Volume 70, pp. 105-110.

Laevers, P., Terryn, H., Vereecken, J. & Thompson, G., E. (1993). , A study of the mechanism of A.C. electrolytic graining of aluminium in hydrocloric and nitric acid, Corrosion Science, Volume 35, pp.231-238

Laevers, P. (1995). Study of the mechanism of the A.C. electrolytic graining of aluminium, VUB, Ph.D. Thesis, Brussels, Belgium

Laevers, P., Terryn, H., Vereecken, J., Kernig, B. & Grzemba, B. (1996). The influence of Manganese on the AC electrolytic graining of aluminium, Corrosion Science, Volume 38, pp. 413-429

Laevers, P., Hubin, A., Terryn, H. & Vereecken, J. (1998). A wall-jet electrode reactor and its application to the study of electrode reaction mechanisms Part III: Study of the mechanism of the AC electrolytic graining of aluminium in hydrochloric acid, Applied Electrochemistry, Volume 28, pp. 387-396

Lee, W., J. & Pyun, S., I. (2000). Effects of sulphate ion additives on the pitting corrosion of pure aluminium in 0.01 M NaCl solution, Electrochimica Acta, Volume 45, pp. 1901–1910

Lin, C., S., Chang,C., C. & Fu H.M. (2001). AC electrograining of aluminum plate in hydrochloric acid, Materials Chemistry and Physics, Volume 68, pp.217–224

Lin, C., S. & Chiu, C., C. (2005). Formation of the Layered Etch Films on AA1050 Aluminum Plates Etched in Nitric Acid Using Alternating Currents, Electrochemical Society, Volume 152, pp. C482-C487

Lin, C., S. & Li, W., J. (2006). Pitting Behavior of Aluminum Foil during Alternating Current Etching in Hydrochloric Acid Containing Sulfate Ions, Electrochemical Society, Volume 153, pp. C51-C56

Lindseth, I. (1999). Optical total reflectance, near-surface microstructure and topography of rolled aluminium materials, NTNU, Ph.D. Thesis, Trondheim, Norway

Marshall, G., J. & Ward, J., A. (1995). Material Science Technology, Volume 11, pp. 1015

Martinez-Caicedo, C., E., Koroleva, E., V., Thompson, G., E., Skeldon, P., Shimizu, K., Habazaki, H. & Hoellrigl, G. (2002). Surface nanotextures on aluminium, Surface and Interface Analysis, Volume 34, pp. 405-408

Meng, G., Wei, L., Zhang, T., Shao, Y., Wang, F., Dong, C. & Li, X. (2009). Effect of microcrystallization on pitting corrosion of pure aluminium , Corrosion Science, Volume 51, pp. 2151-2157

Ono, S. & Habazaki, H. (2009). Effect of sulfuric acid on pit propagation behavior of aluminium under AC etch process, Corrosion Science, Volume 51, pp.2364–2370

Ono, S. & Habazaki, H. (2010). Role of cathodic half-cycle on AC etch process of aluminium, Corrosion Science, Volume 52, pp. 2164-2171

Ono, S. & Habazaki, H. (2011). Pit Growth Behavior of Aluminium under Galvanostatic Control, Corrosion Science, Accepted Manuscript: 22 June 2011

Pourbaix, M. (1974). Atlas of electrochemical equilibria in aqueous solutions, National Association of Corrosion Engineers, Houston, USA

Raes, M. (2009) Internal Report, Vrije Universiteit Brussels, Belgium

Rodriguez, B., Kernig, B., Hasenclever, J. & Terryn, H. (2011) Influence of the surface activation and local pitting susceptibility on the AC-electrograining of aluminium alloys, Corrosion Science, Volume 53, pp. 930–938

Sanchez, S., A., Gonzalez-Garcia, J., Esclapez, M., D., Diez-Garcia, M., I., Del Rio, J., F. & Gazapo, J., L. (2010). Electrograining of aluminium in HCl: effect of the alloy for high-speed processing lines, Surface and Interface Analysis, Volume 42, pp. 311-315

Schaefer, D., W., Shelleman, R., A., Keefer, K., D. & Martin, J., E. (1986). Equilibrium structure and rigidity of alumina polymers, Physica A:Statistical Mechanics and its Applications, Volume 140, pp. 105-113

Smialowska, S., Z. (1986). Pitting Corrosion of Metals, National Association of Corrosion Engineers, Houston , Texas

Smialowska, S., Z. (1999). Pitting Corrosion of Aluminium, Corrosion Science, Volume 41, pp. 1743-1767

Sinko, K., Mezei, R., Rohonczy, J. & Fratzl, P. (1999). Gel Structures Containing Al(III), Langmuir: surfaces and colloids, Volume 15, pp. 6631-6636.

Thompson, G., E., & Wood, G., C. (1978). The effect of alternating voltage on aluminium electrodes in hydrochloric acid, Corrosion Science, Volume 18, pp. 721-746

Terryn, H. (1987). Electrochemical Investigation of AC Electrograining of Aluminium and its porous anodic oxidation, Ph.D. Thesis, Brussels, Belgium

Terryn, H., Vereecken, J. & Thompson, G., E. (1988), Electrograining of aluminum, Transactions of the Institute of Metal Finishing, Volume 66, pp. 116-121

Terryn, H., Vereecken, J. & Thompson, G., E. (1991a), The electrograining of aluminium in Hydrochloric acid- I. Morphological appearance, Corrosion Science, Volume 32, pp. 1159-1172

Terryn, H., Vereecken, J. & Thompson, G., E. (1991b), The electrograining of aluminium in Hydrochloric acid- II. Morphological appearance, Corrosion Science, Volume 32, pp 1173-1188

Tomasoni, F., Van Parys, H., Terryn, H., Hubin, A. & Deconinck, J. (2010). Identification of bubble evolution mechanisms during AC electrograining, Electrochemistry Comminications, Volume 12, pp. 156-159

Tomcsanyi, L., Varga, K., Bartik, I., Horanyi, G. & Maleczki, E. (1989). Electrochemical study of the pitting corrosion of aluminium and its alloys. II. Study of the interaction of chloride ions with a passive film on aluminium and initiation of pitting corrosion, Electrochimica Acta, Volume 34, pp.855–859.

Vargel, C., Jacques, M. & Schmidt, M., P. (2004). Corrosion of Aluminium, Elsevier, pp. 81-605

Wilson, B., P., Dotremont, A., Biesemans, M., Willem, R., Campestrini, P. & Terryn, H. (2008). Effect of Additives on Smut-layer Formation and Pitting during Aluminium Etching in Hydrochloric Acid, Electrochemical Society, Volume 155, pp. C22-C31

Wilson, B. (2006), Internal Report, Vrije Universiteit Brussels, Belgium

7

Corrosive Effects of Chlorides on Metals

Fong-Yuan Ma
Department of Marine Engineering, NTOU
Republic of China (Taiwan)

1. Introduction

1.1 Introduce of pitting corrosion

Alloying metallic elements added during the making of the steel increase corrosion resistance, hardness, or strength. The metals used most commonly as alloying elements in stainless steel include chromium, nickel, and molybdenum. Due to the alloy contains different, in general, stainless steels are separated the grouped into martensitic stainless steels, ferritic stainless steels, austenitic stainless steels, duplex (ferritic-austenitic) stainless steels and precipitation-hardening stainless steels.

Since stainless steel resists corrosion, maintains its strength at high temperatures, and is easily maintained, it is widely used in items such as automotive, propulsion shaft for high speed craft and food processing products, as well as medical and health equipment. The displacement of a high-speed craft is lighter than the displacement of a conventional ship. This displacement aspect is the essential parameter to obtain fast and competitive sea transportations. High-speed craft allows for the use of non-conventional shipbuilding materials provided that a safety standard at least equivalent to that of a conventional ship is achieved. The chapter will describe the corrosion characteristics stainless steels for SUS630,

Stainless steels are used in countless diverse applications for their corrosion resistance. Although stainless steels have extremely good general resistance, stainless steels are nevertheless susceptible to pitting corrosion. This localized dissolution of an oxide-covered metal in specific aggressive environments is one of the most common and catastrophic causes of failure of metallic structures. The pitting process has been described as random, sporadic and stochastic and the prediction of the time and location of events remains extremely difficult. Many contested models of pitting corrosion exist, but one undisputed aspect is that manganese sulphide inclusions play a critical role.

Pitting corrosion is localized accelerated dissolution of metal that occurs as a result of a breakdown of the otherwise protective passive film on metal surface. The phenomenology of pitting corrosion is discussed, including the effects of alloy composition, environment, potential, and temperature. Those have focused on various stages of pitting process, including breakdown of the passive film, metastable pitting, and pit growth. The mechanism of pitting corrosion is similar to that of Crevice corrosion: dissolution of the passivating film and gradual acidification of the electrolyte caused by its insufficient aeration.

Within the pits, an extremely corrosive micro-environment tends to be established, which may bear little resemblance to the bulk corrosive environment. For example, in the pitting of stainless steels in water containing chloride, a micro-environment essentially representing hydrochloric acid may be established within the pits. The pH within the pits tends to be lowered significantly, together with an increase in chloride ion concentration, as a result of the electrochemical pitting mechanism reactions in such systems. Pitting is often found in situations where resistance against general corrosion is conferred by passive surface films. A localized pitting attack is found where these passive films have broken down. Pitting attack induced by microbial activity, such as sulfate reducing bacteria also deserves special mention. Most pitting corrosion in stainless alloys occurs in neutral-to-acid solutions with chloride or ions containing chlorine.

The detection and meaningful monitoring of pitting corrosion usually represents a major challenge. Monitoring pitting corrosion can be further complicated by a distinction between the initiation and propagation phases of pitting processes. The highly sensitive electrochemical noise technique may provide early warming of imminent damage by characteristic in the pit initiation phase. Pitting failures can occur unexpectedly, and with minimal overall metal loss.

Furthermore, the pits may be hidden under surface deposits, and/or corrosion products. A small, narrow pit with minimal overall metal loss can lead to the failure of an entire engineering system. Pitting corrosion, which, for example, is almost a common denominator of all types of localized corrosion attack, may assume different shapes. Corrosion of metals and alloys by pitting constitutes one of the very major failure mechanisms. Pits cause failure through perforation and engender stress corrosion cracks and the life cycle of stainless alloy will decrease.

2. Pitting corrosion phenomenon

The pits often appear to be rather small at the surface, but may have larger cross-section areas deeper inside the metal. Since the attack is small at the surface and may be covered by corrosion products, a pitting attack often remains undiscovered until it causes perforation and leakage. The experimental evaluation of the parameters of a general stochastic model for the initiation of pitting corrosion on stainless steels is described. The variation of these parameters with experimental conditions is used in the development of a microscopic model. A microscopic model which accords with the observed behavior attributes the initiation of pitting corrosion to the production and persistence of gradients of acidity and electrode potential on the scale of the surface roughness of the metal. The observed fluctuations are related to fluctuations in the hydrodynamic boundary layer thickness. A pit becomes stable when it exceeds a critical depth related to the surface roughness.

2.1 Stainless steel metal

In 1913, English metallurgist Harry Brearly, working on a project to improve rifle barrels, accidentally discovered that adding chromium to low carbon steel gives it stain resistance. In addition to iron, carbon, and chromium, modern stainless steel may also contain other elements, such as nickel, niobium, molybdenum, and titanium. Nickel, molybdenum, niobium, and chromium enhance the corrosion resistance of stainless steel. It is the addition of a minimum of 12% chromium to the steel that makes it resist rust, or stain 'less' than other types of steel.

According to the MIL HDBK-73S and Japanese Industry Standard (JIS), there are different stainless steels grades with different corrosion resistance and mechanical properties:

a. Stainless Steel grades 200 Series

This group of alloys is similar to the more common 300 Series alloys described below as they are non-magnetic and have an austenitic structure. The basic Stainless Steel Grades 200 alloy contains 17% chromium, 4% nickel and 7% manganese. These alloys are, however, not immune to attack and are very susceptible to concentration cell corrosion and pitting corrosion attack. When corrosion starts they usually corrode rapidly and non-uniformly. In seawater immersion, the incubation time for these alloys is in the range of 1 to 3 months with some of the Nitronic grades having incubation times of up to 1 year.

b. Stainless Steel Grades 300 Series

This group of alloys are non-magnetic and have an austenitic structure. The basic Stainless Steel Grades 300 alloy contains 18% chromium and 8% nickel. These alloys are subject to crevice corrosion and pitting corrosion. They have a range of incubation times in seawater ranging from essentially zero in the case of the free machining grades, such as Type 303, to 6 months to 1 year for the best alloys, such as Type 316. They have been widely used in facilities with mixed results. If used in an application where chloride levels are low or where concentration cell corrosion has been prevented through design, they are likely to perform well. When chloride levels are high and where concentration cells can occur, the performance of these alloys is often poor. They must always be selected with care for a specific application and the effect of potential non-uniform attack on system performance must be addressed.

c. Stainless Steel Grades 400 Series

This group of alloys are magnetic and have a martensitic structure. The basic alloy contains 11% chromium and 1% manganese. These Stainless Steel Grades alloys can be hardened by heat treatment but have poor resistance to corrosion. They are subject to both uniform and non-uniform attack in seawater. The incubation time for non-uniform attack in chloride containing environments is very short, often only hours or a few days. Unless protected, using these Stainless Steel Grades in sea water or other environments where they are susceptible to corrosion is not recommended.

d. Stainless Steel Grades 600 Series

This series of stainless steels grade is commonly referred to as Precipitation Hardening stainless steels. These steels can be heat treated to high strength levels. They are subject to crevice corrosion and pitting in chloride containing environments and are also subject to stress corrosion cracking and hydrogen embrittlement. The incubation time for crevice corrosion and pitting in seawater is relatively short, often only a few days. The incubation time for stress corrosion cracking can be very short, sometimes measured in hours.

The use of this Stainless Steel Grade in chloride containing environments is not normally recommended unless they are carefully selected, their heat treatment is carefully specified and controlled, and the effect of pitting and crevice corrosion is properly addressed.

2.2 Principle of pitting corrosion

Pitting corrosion is an electrochemical oxidation-reduction process, which occurs within localized deeps on the surface of metals coated with a passive film.

Anodic reactions inside the pit:

$$Fe = Fe^{2+} + 2e^{-} \text{ (dissolution of iron)}$$

The electrons given up by the anode flow to the cathode where they are discharged in the cathodic reaction:

$$1/2O_2 + H_2O + 2e^{-} = 2(OH^{-})$$

As a result of these reactions the electrolyte enclosed in the pit gains positive electrical charge in contrast to the electrolyte surrounding the pit, which becomes negatively charged.

The positively charged pit attracts negative ions of chlorine Cl^{-} increasing acidity of the electrolyte according to the reaction:

$$FeCl_2 + 2H_2O = Fe(OH)_2 + 2HCl$$

pH of the electrolyte inside the pit decreases from 6 to 2-3, which causes further acceleration of corrosion process.

Large ratio between the anode and cathode areas favors increase of the corrosion rate. Corrosion products ($Fe(OH)_3$) form around the pit resulting in further separation of its electrolyte.

2.3 Stages of pitting corrosion

Pitting corrosion is treated as a time-dependent stochastic damage process characterized by an exponential or logarithmic pit growth. Data from propulsion shaft for high-speed craft is used to simulate the sample functions of pit growth on metal surfaces. Perforation occurs when the deepest pit extends through the thickness of the propulsion shaft. Because the growth of the deepest pit is of stochastic nature, the time-to-perforation is modeled as a random variable that can be characterized by a suitable reliability model. It is assumed that corrosion will occur at multiple pits on both sides of the rivet deeps and will cause multiple fatigue cracks. Therefore, system failure could occur due to the linkage between any two neighboring cracks.

2.3.1 Pit initiation

An initial pit may form on the surface covered by a passive oxide film as a result of the following:

a. Mechanical damage of the passive film was caused by scratches. Anodic reaction starts on the metal surface exposed to the electrolyte. The passivity surrounding surface is act as the cathode.
b. Particles of a second phase emerging on the metal surface. These particles precipitating along the grains boundaries may function as local anodes causing localized galvanic corrosion and formation of initial pits.
c. Localized stresses in form of dislocations emerging on the surface may become anodes and initiate pits.
d. Non-homogeneous environment may dissolve the passive film at certain locations where initial pits form.

2.3.2 Pitting growth

In presence of chloride ions pits are growing by autocatalytic mechanism. Pitting corrosion of a stainless steel is illustrated in the Figure 1. The actual pitting corrosion phenomenon is shown on propeller shaft of high speed craft, and the pit depth was measured with dial gauge as shown in Figure 2.

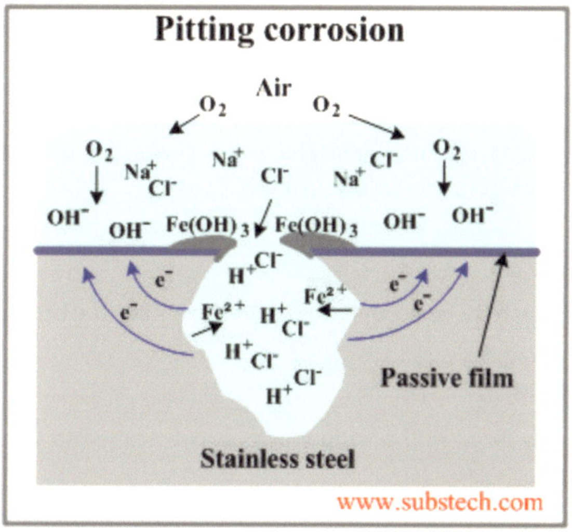

Fig. 1. Pitting corrosion deep growth

Fig. 2. Measuring depth of pitting deep

2.3.3 Transition from pitting to fatigue crack nucleation

The third stage is the transition from pit growth to fatigue crack nucleation, where mechanical effects such as the stress intensity factor come into play. The nucleation of the corrosion crack is essentially a competition between the processes of pit growth and crack growth. Two criteria are used to describe the transition process.

2.3.4 Short crack growth

The short crack growth stage involves chemical and microstructural factors and their interactions. Although much research has been done in this area, it has been difficult to derive an explicit formula for short crack growth, especially in corrosive environment. For computational simplicity, a probabilistic power law model is presented here to describe the relationship between the stress intensity factor and the growth rate. In this method, an empirically based probabilistic relationship is used to model the corrosion short crack growth.

2.3.5 Transition from short crack growth to long crack growth

Experimental and analytical approaches have been proposed to determine the transition size from short crack growth to long crack growth.

2.3.6 Long crack growth

The widely used Paris Law may be used in this stage to estimate the time for long crack growth.

2.3.7 Crack coalescence

The linkage between any two neighboring cracks is considered to be the failure criterion at this stage.

2.4 Pitting corrosion behavior

Corrosion of metals and alloys by pitting constitutes one of the very major failure mechanisms. Pits cause failure through perforation and engender stress corrosion cracks. Pitting is a failure mode common to very many metals. It is generally associated with particular anions in solution, notably the chloride ion. The origin of pitting is small. Pits are nucleated at the microscopic scale and below. Detection of the earliest stages of pitting requires techniques that measure tiny events.

Stainless steels are used in countless diverse applications for their corrosion resistance. Although stainless steels have extremely good general resistance, stainless steels are nevertheless susceptible to pitting corrosion. This localized dissolution of an oxide-covered metal in specific aggressive environments is one of the most common and catastrophic causes of failure of metallic structures. The pitting process has been described as random, sporadic and stochastic and the prediction of the time and location of events remains extremely difficult. Many contested models of pitting corrosion exist, but one undisputed aspect is that manganese sulphide inclusions play a critical role.

The chromium in the steel combines with oxygen in the atmosphere to form a thin, invisible layer of chrome-containing oxide, called the passive film. The sizes of chromium atoms and their oxides are similar, so they pack neatly together on the surface of the metal, forming a stable layer only a few atoms thick. If the metal is cut or scratched and the passive film is disrupted, more oxide will quickly form and recover the exposed surface, protecting it from oxidative corrosion.

The passive film requires oxygen to self-repair, so stainless steels have poor corrosion resistance in low-oxygen and poor circulation environments. In seawater, chlorides from the salt will attack and destroy the passive film more quickly than it can be repaired in a low oxygen environment.

Some metals show preferential sites of pit nucleation with metallurgical microstructural and microcompositional features defining the susceptibility. However, this is not the phenomenological origin of pitting per se, since site specificity is characteristic only of some metals. A discussion is presented of mechanisms of nucleation; it is shown that the events are microscopically violent. The ability of a nucleated event to survive a series of stages that it must go through in order to achieve stability is discussed. Nucleated pits that do not propagate must repassivate. However, there are several states of propagation, each with a finite survival probability. Several variables contribute to this survival probability.

2.5 Electron fractography of fatigue fracture with pitting corrosion

The evolution of corrosion pits on stainless steel immersed in chloride solution occurs in three distinct stages: nucleation, metastable growth and stable growth. A microcrack generated by pitting corrosion, forms the initial origin for fatigue fracture. But in fact, fatigue failure is not certainly caused at the deepest interior pitting deep. Owing to the different shape of interior pitting deeps as shown in Figure 3, some of them are hard to start or to continue the crack propagation and play as a role of crack arrester, as shown in Figure 4. If the fracture is caused by the pitting deep as an inclusion, then continuous plastic deformation can be found around the pitting deep. The dimples are generally equalized by overloaded tension, and elongated by shear or tearing, as shown in Figure 5 and Figure 6.

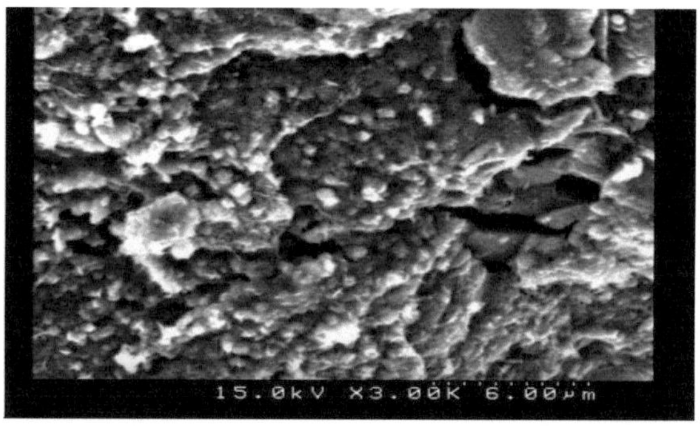

Fig. 3. Interior pitting deep

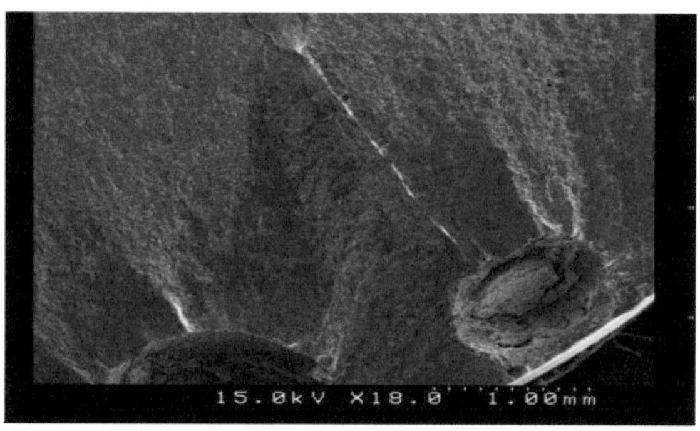

Fig. 4. Fatigue failure arrester at interior pitting deep

Fig. 5. Equalized dimples around interior pitting deep

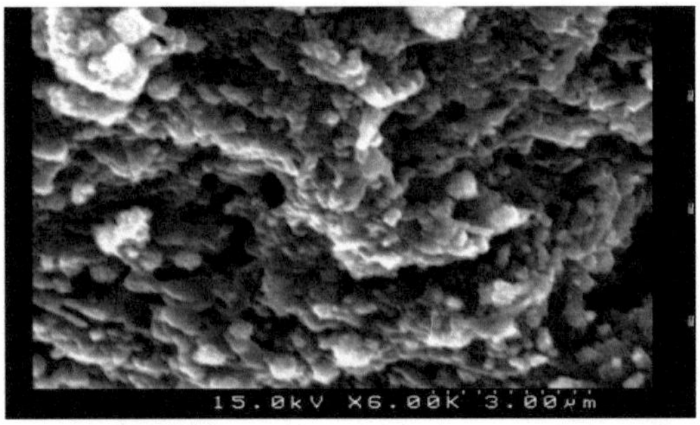

Fig. 6. Elongated dimples around interior pitting deep

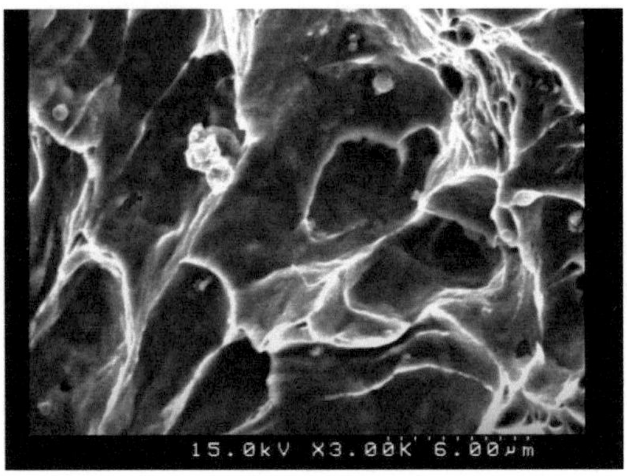

Fig. 7. Typical cleavage fractures in fracture area (3000 times)

The features of cleavage failure can be seen by flat fractography as shown in Figure 7 and Figure 8 with 3000 and 1000 times Scanning Electron Microscopy (SEM) respectively. Cleavage failure occurs by separation along crystallographic planes. This transgranular fracture is categorized to be the brittle fracture in the fracture area. There are several features can be identified to be cleavage, namely, herringbone, tongues and river stream. Figure 9 and Figure 10 show the microscopic fractographics of river stream and tongue patterns, which have seen in the study.

At grain boundaries the fracture plane or cleavage plane changes because of the differences of crystallographic orientations. Cleavages are not only associated with transgranular fracture, but also with brittle particles as shown in Figure 11 and Figure 12 respectively.

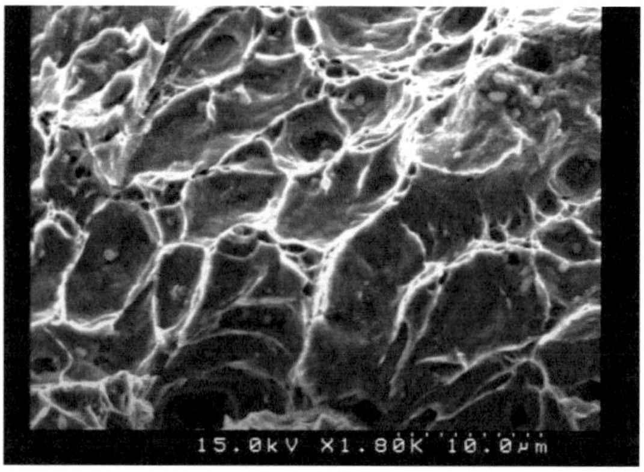

Fig. 8. Typical cleavage fractures in fracture area (1000 times)

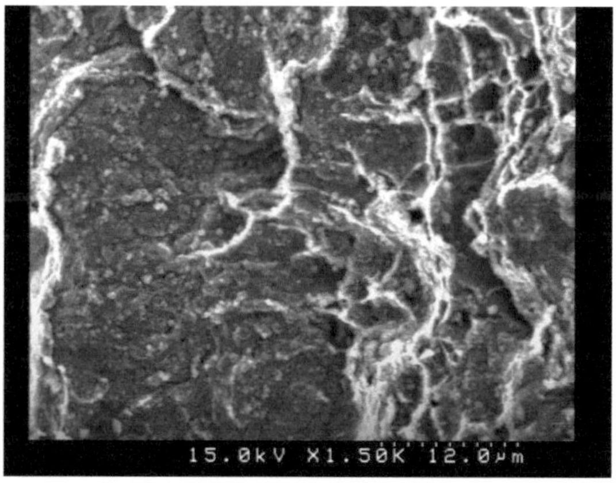

Fig. 9. River pattern in fatigue propagation area (1500 times)

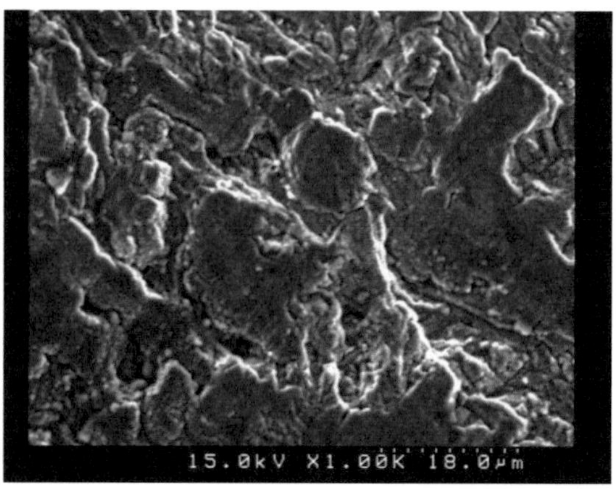

Fig. 10. Tongue pattern in fatigue propagation area (1000 times)

Some metals can fail in brittle manner, but do not cleave. These fractures are identified as quasi-cleavage. They are similar to cleavage but their features are usually fairly flat and smaller, as shown in Figure 13 and Figure 14.

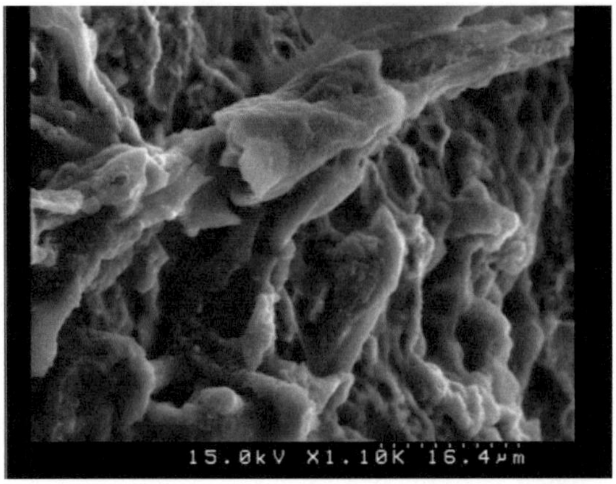

Fig. 11. Cleavage with transgranular fracture in fatigue propagation area

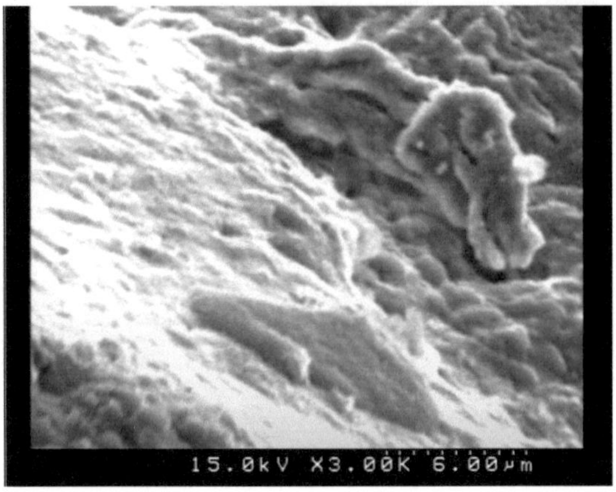

Fig. 12. Cleavage with brittle particles in fracture area

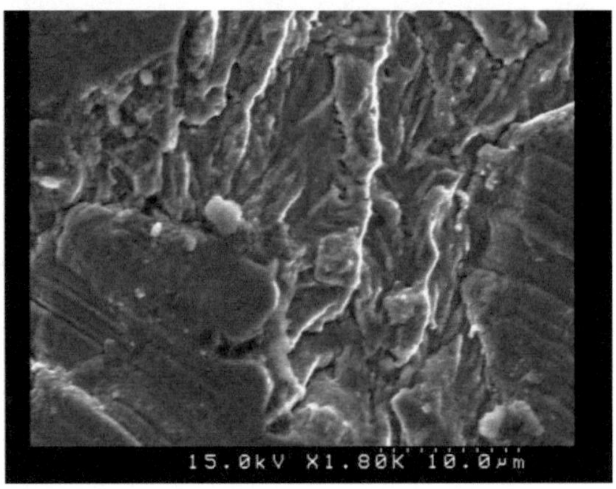

Fig. 13. Quasi-cleavage in fracture area (I)

Also the intergranular fracture can occur by a number of causes, but it is generally possible to be identified fractographically by the features of grain contours, grain boundaries and triple points. Examples of this kind in tergranular fractures are tree pattern fracture and sub-crack caused by the interior pitting deeps, as shown as Figure 15 and Figure 16.

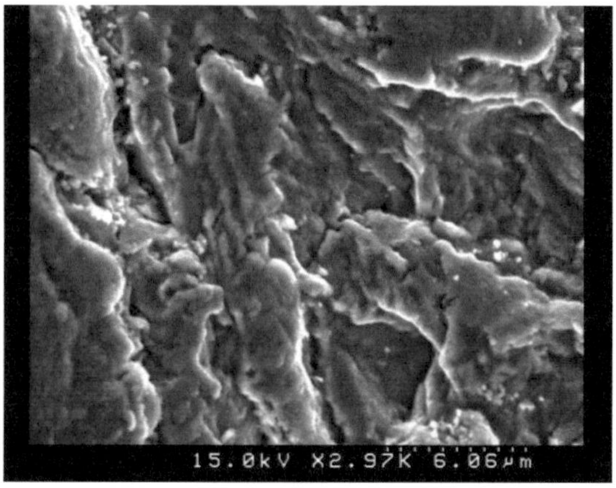

Fig. 14. Quasi-cleavage in fracture area (II)

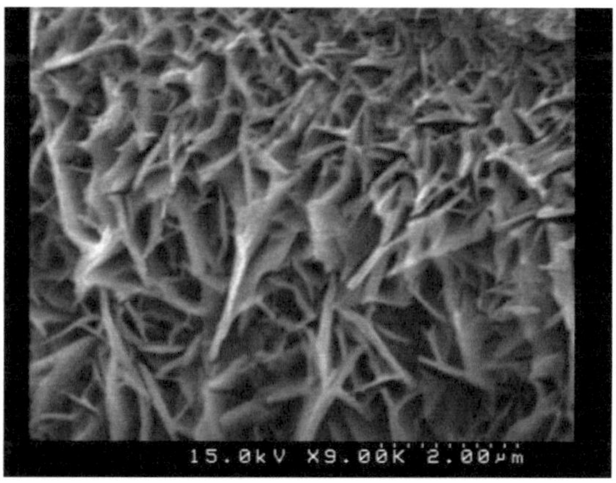

Fig. 15. Tree pattern fracture caused by interior pitting deep

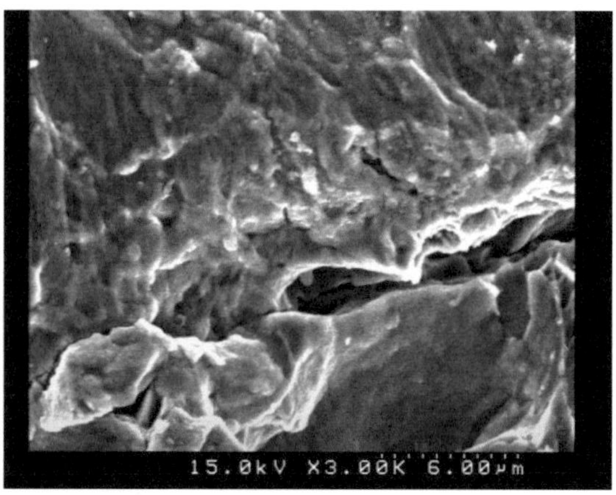

Fig. 16. Sub-crack caused by interior pitting deep

3. Initial pitting deep and pitting corrosion behavior

Pitting corrosion is a localized form of corrosion by which cavities or deeps are produced in the material. Pitting is considered to be more dangerous than uniform corrosion damage because it is more difficult to detect, predict and design against. Corrosion products often cover the pits. A small, narrow pit with minimal overall metal loss can lead to the failure of an entire engineering system.

3.1 Corrosion pit shapes

Pitting corrosion forms on passive metals and alloys like stainless steel when the ultra-thin passive film is chemically or mechanically damaged and does not immediately re-passivity. The resulting pits can become wide and shallow or narrow and deep which can rapidly perforate the wall thickness of a metal.

3.1.1 Pitting shape by ASTM

Pitting corrosion can produce pits with their mouth open or covered with a semi-permeable membrane of corrosion products. Pits can be either hemispherical or cup-shaped. In some cases they are flat-walled, revealing the crystal structure of the metal, or they may have a completely irregular shape. Pitting corrosion occurs when discrete areas of a material undergo rapid attack while most of the adjacent surface remains virtually unaffected.

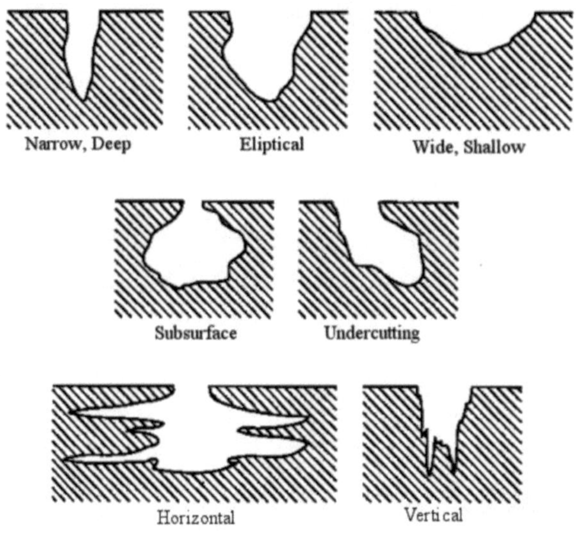

Fig. 17. ASTM-G46 has a standard visual chart for rating of pitting corrosion (http://www.corrosionclinic.com)

3.1.2 Experiment pitting shape

Localized chemical or mechanical damage to the protective oxide film; water chemistry factors which can cause breakdown of a passive film are acidity, low dissolved oxygen concentrations and high concentrations of chloride in seawater. The actual pitting shape was investigated by Scanning Electron Microscopy shown as Figure 18 and Figure 19.

Fig. 18. Pitting deep shape

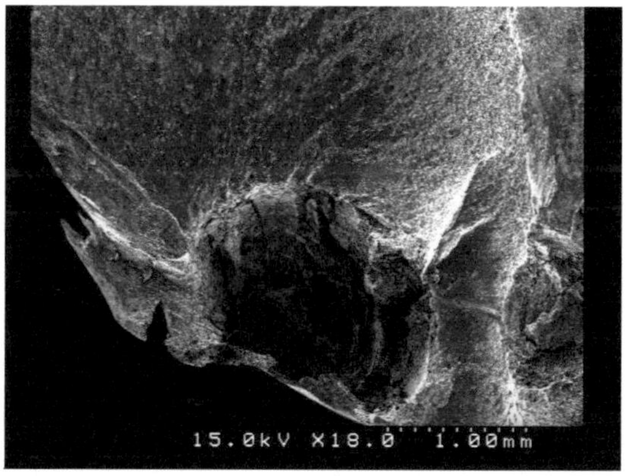

Fig. 19. Pitting deep shape

3.2 Fatigue initial point

In this section, the standard of material specimen is based on JIS G4303, 36 pieces specimen have been prepared for test and separated 4 groups, each group include 3 sets and 1 set include 3 pieces. Firstly, the acceleration pitting corrosion test has been carried out for all sampling pieces. After pitting corrosion, that be separated A, B, C and D group based on the post treatment condition. A group is without post treatment; B group is with solution treatment; C group is with preheated 1400℃ for metal coating and D group is preheated 200℃ for metal coating. The metal coating surface had been machined two sets among and other one set without machined for C and D group. The effects of the morphology and matrix structure on the fatigue process as well as fatigue life were examined through metallurgical graphic and fracture graphic observation. The metallurgical graphs of the fatigue fracture surfaces were studied by SEM and are shown in Figures 20 to Figure 23.

Under all test conditions, Figure 20 the crack initiation of Group A, was found to occur at the cavity inside the pitting deep; Figure 21, the crack initiation of Group B, the diameter of pitting deep have clearly increased from 0.1 mm to 0.5 or 1.0mm after solution treatment, and cannot find the fatigue prorogation from the fracture surface. So, the fatigue life cycles had obviously decreased to compare between Table 4 and Table 5; Figure 22 6, the crack initiation of Group C, the microstructure of initiation point have different change with Group A due to the preheated process and can be found to occur at the fracture in the boundary; Figure 23, the crack initiation of Group D, the microstructure of initiation point have cleavage and fracture phenomenon. The tests without the metal surface, cannot find the fatigue initiation point and the fracture point in the near middle of surface, and the fatigue life cycle longer than other specimen.

The observations indicated that the fatigue life is not influenced in significant way by the mechanisms of interaction between dislocations and precipitates. It has been reported that at room temperature the fatigue life is influenced by the deformation mechanism. We believe that in the present work, which has been carried out at elevated temperature and under laboratory atmosphere, the fatigue life is essentially determined by the cracks initiated at the specimen surface as a result of intercrystal line oxidation. This view is supported by the observations, which are reported in Figure.

Fig. 20. Fatigue initial point of specimen (Group A)

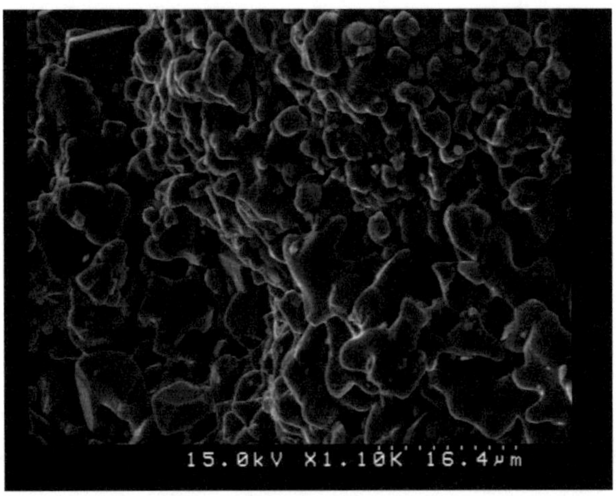

Fig. 21. Fatigue initial point of specimen (Group C)

Fig. 22. Fatigue initial point of specimen (Group B)

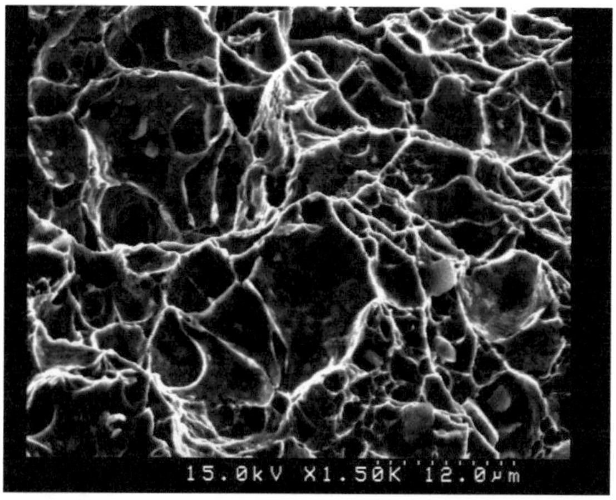

Fig. 23. Fatigue initial point of specimen (Group D)

3.3 Fatigue fracture mechanism observed by SEM

From the electron fractograhpy of the test specimen, there appears many different kinds of characteristic fatigue fracture features, which encompass the fatigue lines, the fatigue beach marks, the fatigue stair line, the critical crack length and the fracture area, etc.

3.3.1 Fatigue lines

In the fatigue crack propagation process, the fatigue lines appear along and in front of the trajectory of macroscopic plastic deformation at different instant, as shown in Figure 24. Usually, the normal direction of these fatigue lines is referred to as the direction of fatigue crack propagation.

3.3.2 Fatigue beach marks

Beach marks are similar to the fatigue lines that occur before fatigue crack and along the route of plastic deformation traced, as shown in Figure 25. Usually, their normal direction approximately aims at the direction of the fatigue crack propagating. The number and the thickness of beach marks can be considered as the basis to measure fatigue endurance.

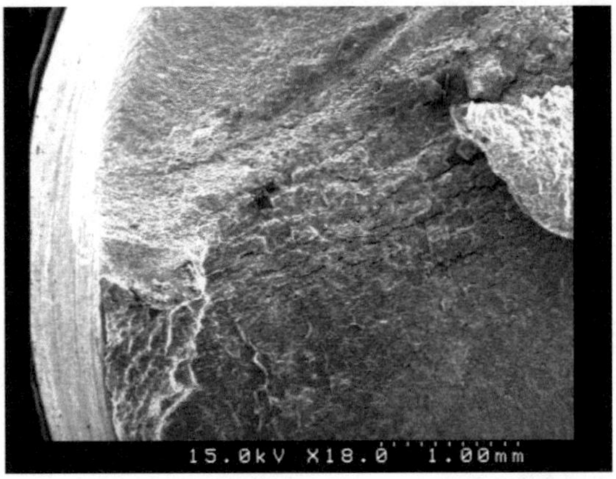

Fig. 24. Fatigue lines in fatigue fracture surface

Fig. 25. Beach marks in initial fatigue stage

3.3.3 Fatigue stair lines

In various situations of fatigue failures, the fatigue stair line is one of the main characteristics of fatigue fracture. The fatigue stair lines can be often detected in the zone of obvious bright-gloomy band or curves, as shown in Figure 26. They are resulted from the merging or intersection of different fatigue crack propagation generated at different time and origins. Generally the direction of fatigue stair lines is normal to the direction of fatigue crack propagation. The direction and the density of the fatigue stair lines are the characteristic parameters used to identify the locations of fatigue origin sources.

3.3.4 Critical crack length

Critical (fatigue) crack length means the limit length of origin crack at which the fatigue crack propagation can occur. Beyond the critical crack length of origin crack source, it represents that the fracture occurs abruptly. Such critical fatigue crack length can be seen in Figure 27.

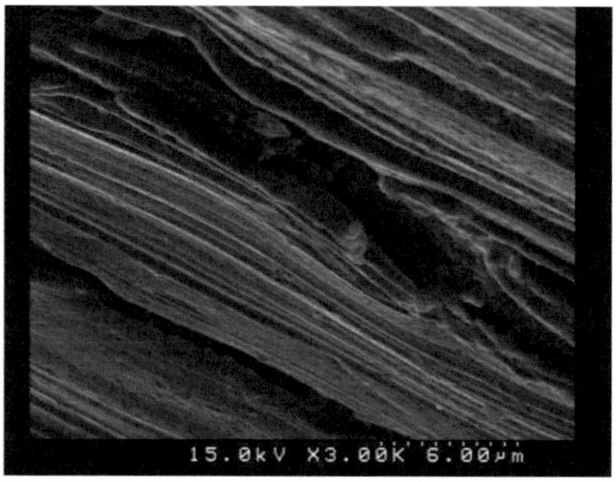

Fig. 26. Fatigue stair lines

Fig. 27. Critical crack length

3.3.5 Fracture area

In the instantaneous fracture zone or referred to as the fracture area, it can be seen some crystalline patterns like river stream and tongue as shown in Figure 28 and Figure 29.

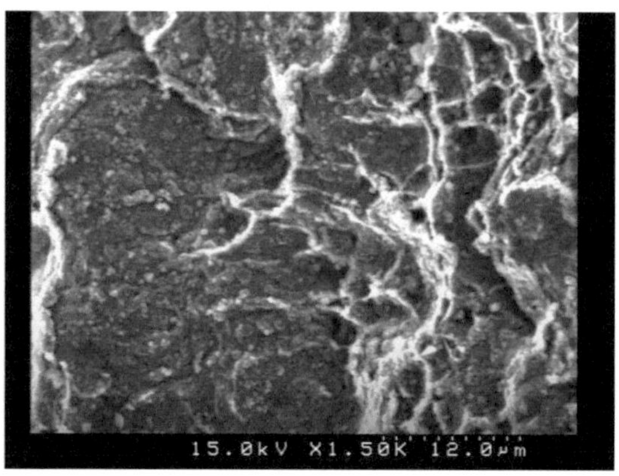

Fig. 28. Fracture area with river stream pattern

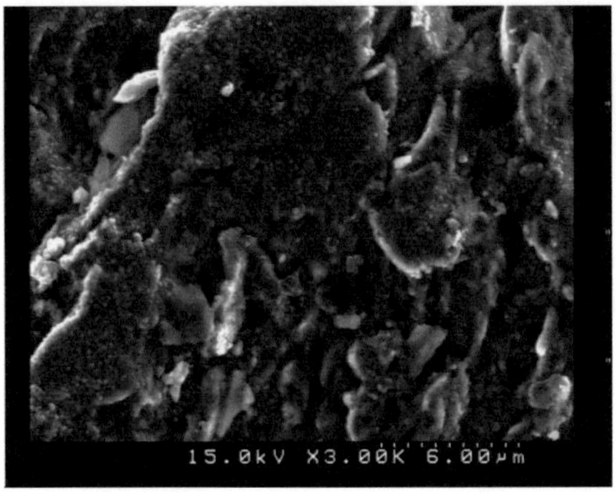

Fig. 29. Fracture area with tongue pattern

3.3.6 Transgranular and intergranular fracture

In the fracture area, fractures can be categorized as transgranular and intergranular fracture based on the fracture path related to the crystalline grain structure. Transgranular fracture may be ductile or brittle depending on the amount of grain deformation as shown in Figure 30. While intergranular fracture is only pertaining to a brittle one, since the separation of grains is usually due to the presence of brittle interface, as shown in Figure 31.

4. Pitting corroding rate

The grey systems are the systems that lack information, such as architecture, parameters, operation mechanism and system behavior, for example, for estimating the tendency of Typhoon landing in Taiwan [8]. There are a number of factors that affect the pitting corrosion rate of stainless alloy that include [a, b, c… and x] as previously mentioned.

The grey correlation analysis explains uncertain correlations between one main factor and all other factors in a given system. The grey correlation analysis method is based on the clustering approach in which the time factor is during the experiment period.

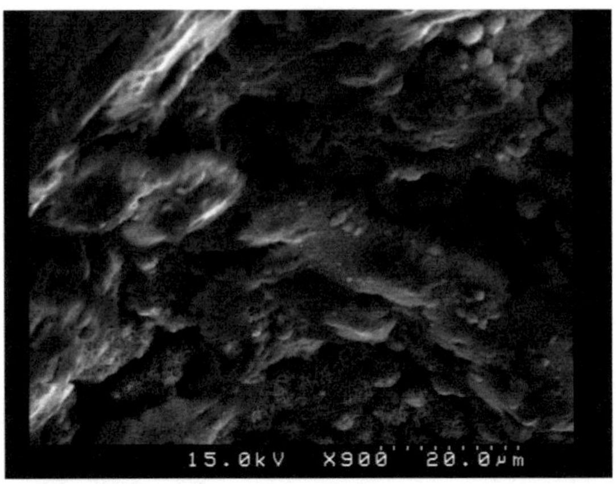

Fig. 30. Transgranular fracture at interior pitting deep

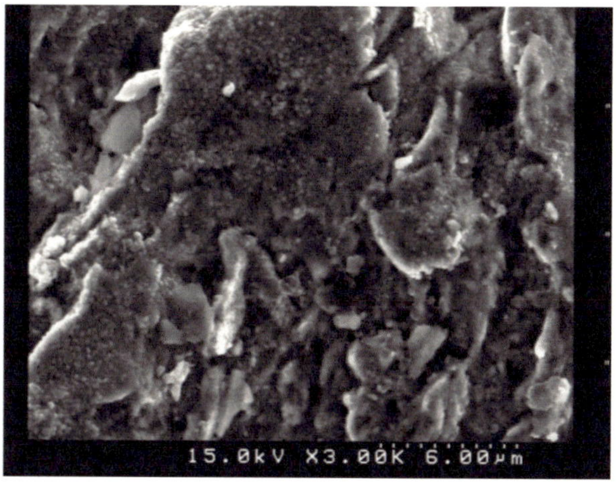

Fig. 31. Intergranular fracture in the fracture area

4.1 Material and experiment procedure

For this test, stainless SUS 630 was used and test specimen with a size of 50(L)*25(B)*5(t) mm. The Chemical property is shown as in Table 3.1 and the mechanical property is as shown in Table 3.2 for the SUS 630. All specimens were carried out the pitting corrosion test separated for the short term test (acceleration pitting corrosion test) and for long term test (environmental test). The acceleration pitting corrosion test was separated six (6) cases test and the environmental test was separated the static test and dynamic test.

The size of test period relates to size and development trend of the corroding rate. The test period of three chlorine iron solution is 72 hours according to the standard, this test, in order to further understand and corrode the relation with test period, and divided into 24, 48, 72 and 96 hours to wait for four cycles.

This test is divided into six (6) schemes, is mainly for the same test temperature, different pH value, such as Case 1, Case3 and Case 4; Another the same experimental pH value, different test temperature, such as Case 5 and Case 6. Case 2 utilizes Case 1 to finish the solution of the test, because the thickness of its chlorine ion is in order to reduce, repeat another test; it is shown in Table 1 and in detail picture as following specimen picture.

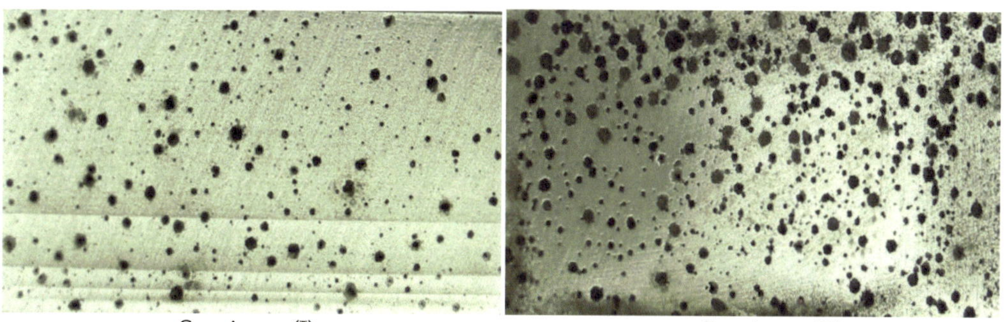

Specimen (I) Specimen (II)

Fig. 32. Specimen Picture

Material							SUS630					
Sample size							50mm*25mm*5mm					
CASE	1		2		3		4		5		6	
Test date	2002/2/14		2002/2/27		2002/3/25		2002/5/07		2002/07/02		2002/07/02	
Test temperature	25 °C		22 °C		25 °C		25 °C		32 °C		26°C	
PH value	1.3		1.2		1.5		1.4		0.8		0.8	
sample groups	A	B	A	B	A	B	A	B	A	B	A	B
Test period	Specimen No.											
24 Hour	1A1	1B1	2A1	2B1	3A1	3B1	4A1	4B1	5A1	5B1	6A1	6B1
48 Hour	1A2	1B2	2A2	2B2	3A2	3B2	4A2	4B2	5A2	5B2	6A2	6B2
72 Hour	1A3	1B3	2A3	2B3	3A3	3B3	4A3	4B3	5A3	5B3	6A3	6B3
96 Hour	1A4	1B4	2A4	2B4	3A4	3B4	4A4	4B4	5A4	5B4	6A4	6B4

Table 1. Testing procedure

Measuring the pitting depth from experiment specimen, it note down ten(10) pitting deep relatively, asks its average, its unit is mm. and then called the corroding rate with the average divided by test period, its unit is mm/h, and about 10^6 times of the standard unit. The result record is shown as the Table 2.

Specimen No.	1A1	1A2	1A3	1A4	1B1	1B2	1B3	1B4
Average	0.352	0.509	1.116	0.945	0.48	0.762	1.04	0.997
Rate of Corrosion	0.0146667	0.0106042	0.0155000	0.0098438	0.0200000	0.0158750	0.0144444	0.0103854
Specimen No.	2A1	2A2	2A3	2A4	2B1	2B2	2B3	2B4
Average	0.00	0.678	0.00	0.785	0.38	0.518	0.55	0.35
Rate of Corrosion	0.0000000	0.0141250	0.0000000	0.0081771	0.0158333	0.0107917	0.0076389	0.0036458
Specimen No.	3A1	3A2	3A3	3A4	3B1	3B2	3B3	3B4
Average	0.399	0.725	0.643	0.763	0.545	0.776	1.024	0.779
Rate of Corrosion	0.0166250	0.0151042	0.0089306	0.0079479	0.0227083	0.0161667	0.0142222	0.0081146
Specimen No.	4A1	4A2	4A3	4A4	4B1	4B2	4B3	4B4
Average	0.441	0.684	0.774	0.908	0.555	0.549	1.025	1.09
Rate of Corrosion	0.0183750	0.0142500	0.0107500	0.0094583	0.0231250	0.0114375	0.0142361	0.0113542
Specimen No.	5A1	5A2	5A3	5A4	5B1	5B2	5B3	5B4
Average	0.654	0.567	0.919	0.934	0.665	1.15	0.91	1.228
Rate of Corrosion	0.0272500	0.0118125	0.0127639	0.0097292	0.0277083	0.0239583	0.0126389	0.0127917
Specimen No.	6A1	6A2	6A3	6A4	6B1	6B2	6B3	6B4
Average	0.384	0.503	0.675	0.702	0.52	0.69	0.736	0.798
Rate of Corrosion	0.0160000	0.0104792	0.0093750	0.0073125	0.0216667	0.0143750	0.0102222	0.0083125

Table 2. Pitting depth (unit mm) and corrode Rate (Unit mm/h)

Environmental test, its purpose in real environment, service life when used for and estimated and moved forward the axle department actually mainly, the two sets have been put in the sea side (Dynamic Test) at Keelung shipyard (China Ship Building Corporation) since May 2004 and four other sets have been put in plastic barrel with sea water (Static Test) in room since August 2004. The average temperature of sea water is around 25°C and the pH value is about 7.2 for this test condition. All test pieces were retrieved monthly and the depth of pitting corrosion were measured and recorded accordingly.

The result record is shown as the Table 3 for the dynamic experiment, its unit is 1/100 mm, and the static experiment record is as shown the Table 4. If the unit was selected to mm/100/day, the Table 3 will be changed to Table 5 and Table 4 also changed to Table 6.

Specimen No.	Jun	Jul	Aug	Sept	Oct	Nov	Dec
SP1	37.5	70	100	125	135	140	147
SP2	36.5	65	80	98	110	125	137
Average	37	67.5	90	111.5	122.5	132.5	142
Specimen No.	Jan	Feb	Mar	Apr	May	Jun	Jul
SP1	152	152	153	154	154.5	155	155.5
SP2	148	151.5	152.5	153	153.5	154	154.5
Average	150	151.75	152.75	153.5	154	154.5	155

Table 3. The result of pitting corrosion depth of dynamic experiment (unit: 1/100mm)

Specimen No.	Sept	Oct	Nov	Dec	Jan	Feb
1SA	35	68	85	98	119	120
2SA	34.5	65	89	96	108	114
3SA	33.5	64	88	97	110	111
4SA	33	60	82	95	103	116
AVG	34	64.25	86	96.5	110	115.25
Specimen No.	Mar	Apr	May	Jun	Jul	
1SA	123	124	125	126	126.5	
2SA	118	120	122	123	123.5	
3SA	114	115.5	117	118	118.5	
4SA	120.5	122.5	123.5	125	125	
AVG	118.875	120.5	121.875	123	123.375	

Table 4. The result of pitting corrosion depth of static experiment (unit: 1/100mm)

Specimen No.	Jun	Jul	Aug	Sept	Oct	Nov	Dec
SP1	1.2500	1.1667	1.1111	1.0417	0.9000	0.7778	0.7000
SP2	1.2167	1.0833	0.8889	0.8167	0.7333	0.6944	0.6524
Average	1.2333	1.1250	1.0000	0.9292	0.8167	0.7361	0.6762
Specimen No.	Jan	Feb	Mar	Apr	May	Jun	Jul
SP1	0.6333	0.5630	0.5100	0.4667	0.4292	0.3974	0.3702
SP2	0.6167	0.5611	0.5083	0.4636	0.4264	0.3949	0.3679
Average	0.6250	0.5620	0.5092	0.4652	0.4278	0.3962	0.3690

Table 5. The result pitting corrosion rate of dynamic experiment (unit. mm/100/day)

Specimen No.	Sept	Oct	Nov	Dec	Jan	Feb
1SA	1.1667	1.1333	0.9444	0.8167	0.7933	0.6667
2SA	1.1500	1.0833	0.9889	0.8000	0.7200	0.6333
3SA	1.1167	1.0667	0.9778	0.8083	0.7333	0.6167
4SA	1.1000	1.0000	0.9111	0.7917	0.6867	0.6444
AVG	1.1333	1.0708	0.9556	0.8042	0.7333	0.6403

Specimen No.	Mar	Apr	May	Jun	Jul	
1SA	0.5857	0.5167	0.4630	0.4200	0.3833	
2SA	0.5619	0.5000	0.4519	0.4100	0.3742	
3SA	0.5429	0.4813	0.4333	0.3933	0.3591	
4SA	0.5738	0.5104	0.4574	0.4167	0.3788	
AVG	0.5661	0.5021	0.4514	0.4100	0.3739	

Table 6. The result pitting corrosion rate of static experiment (unit. mm/100/day)

4.2 Influence factor of corroding rate

According to measured data of pitting deep depth, utilizing the analytical method, can be estimated the distribution. Normal distribution can be regarded as the probability distribution of the corroding rate of stainless steel for pitting corrosion.

T-test is most commonly applied when the test statistic would follow a normal distribution if the value of a scaling term in the test statistic were known. The *t*-distribution is symmetric and bell-shaped, like the normal distribution, but has heavier tails, meaning that it is more prone to producing values that fall far from its mean. Using the Table 2, the upper and lower limit will be calculated by the t-test and t-distribution table shown as Table 7.

CASE 1						
Probability	0.5	0.95	0.99	0.5	0.95	0.99
t value	0.765	3.182	5.841	0.765	3.182	5.841
Lower Limit	0.01156615	0.00813025	0.00435033	0.01257010	0.00433614	-0.00472223
Upper Limit	0.01374113	0.01717703	0.02095695	0.01778232	0.02601628	0.03507466
CASE 2						
Lower Limit	0.00294363	-0.0053717	-0.0145197	0.00752982	0.00137642	-0.00539307
Upper Limit	0.00820741	0.01652281	0.02567078	0.01142503	0.01757843	0.02434793
CASE 3						
Lower Limit	0.01048792	0.00523060	-0.00055312	0.01256486	0.00391395	-0.00560312
Upper Limit	0.01381589	0.01907321	0.02485692	0.01804103	0.02669194	0.03620902
CASE 4						
Lower Limit	0.01168016	0.00685193	0.00154028	0.01252271	0.00457512	-0.00416820
Upper Limit	0.01473650	0.01956473	0.02487638	0.01755367	0.02550126	0.03424459
CASE 5						
Lower Limit	0.01232570	0.00264765	-0.0079994	0.01654468	0.00792052	-0.00156712
Upper Limit	0.01845207	0.02813012	0.03877718	0.02200392	0.03062808	0.04011573
CASE 6						
Lower Limit	0.01232570	0.00264765	-0.0079994	0.01654468	0.00792052	-0.00156712
Upper Limit	0.01845207	0.02813012	0.03877718	0.02200392	0.03062808	0.04011573

Table 7. Pitting deep upper and lower limits (unit mm) of depth

F-test is any statistical in which the test static has an F-distribution under the null hypothesis. It is most often used when comparing statistical models that have been fit to a data set, in order to identify the model that best fits the population from which the data were sampled. Exact *F-tests* mainly arise when the models have been fit to the data using least squares.

The hypotheses that the means of several normally distributed populations, all having the same standard deviation, are equal. This is perhaps the best-known F-test, and plays an important role in the analysis of variance (ANOVA).

In statics, analysis of variance is a collection of statistical models, and their associated procedures, in which the observed variance in a particular variable is partitioned into components attributable to different sources of variation. In its simplest form ANOVA provides a statistical test of whether or not the means of several groups are all equal, and therefore generalizes-test to more than two groups.

For a defect-free material, pitting corrosion is caused by the chemistry that may contain aggressive chemical species such as chloride. Chloride is particularly damaging to the passive film oxide so pitting can initiate at oxide breaks. The influence factors of pitting corrosion for stainless steel were complex, for example, material defect, solution pH value and temperature etc.

4.2.1 The corrodes rate between pitting depth and material defect

a. Under the conditions of 95% confidence interval, its upper and lower limit value are 0.00989783 and -0.00485264 for Case 1, 0.01245305 and -0.00464923 for Case 2, 0.01151913 and -0.00521705 for Case 3, 0.0095169 and -0.00585718 for Case 4, 0.01460108 and -0.00683025 for Case 5, 0.00966449 and - 0.00395963 for Case 6. Based on the above-mentioned value, the intersection of extreme value and it's including 0, According to the t-test principle, there is no obvious relation between corroded rate and material defect.
b. Using the ANOVA methods calculation and F value to assay, the calculated F value is less than the critical value of F value. So, there is no obvious relation between corroded rate and material defect.

4.2.2 The corrodes rate between pitting depth and pH value of solution

a. Under the conditions of 95% confidence interval, its upper and lower limit are 0.0056934 and -0.00468993 for Case 1 and Case 3 Group A, and 0.00999922 and -0.00974575 for Case 1 and Case 3 Group B, Based on the above-mentioned value, the intersection of extreme value and it's including 0, According to the t-test principle, there is no obvious relation between corroded rate and pH value.
b. Using the ANOVA methods calculation and F value to assay, the calculated F value is less than the critical value of F value. So, there is no obvious relation between corroded rate and pH value.

4.2.4 The corrodes rate between pitting depth and solution temperature

a. Under the conditions of 95% confidence interval, its upper and lower limit are 0.01341501 and -0.00468993 for Case 5 and Case 6 Group A, and 0.01476674 and -0.00350632 for Case 5 and Case 6 Group B, Based on the above-mentioned value, the intersection of extreme value and it's including 0, According to the t-test principle, there is no obvious relation between corroded rate and temperature.
b. Using the ANOVA methods calculation and F value to assay, the calculated F value is less than the critical value of F value. So, there is not an obvious relation between corroded rate and temperature.

4.3 Corroding rate

Theoretically, a local cell that leads to the initiation of a pit can be caused by an abnormal anodic site surrounded by normal surface, which acts as a cathode, or by the presence of an abnormal catholic site surrounded by a normal surface in which a pit will have disappeared due to corrosion. The corroding rate was calculated by the grey theory method and regression method and the result was shown as Figure 33 to Figure 38 using the environmental test data.

The precision of the prediction is around 96.44% and error rate is about 3.56% on average with the grey predication for dynamic test; the precision of the prediction is around 96.97% and error rate is about 3.03% on average with the grey predication for static test. Therefore, the grey model GM (1,1) is an acceptable method for the prediction of the Pitting corrosion depth and rate.

The corrosion depth and rate precision of the prediction is individually around 89.84% and 98.43% and error rate is about 10.16% and 1.57% on average with the regression predication for dynamic test; the corrosion depth and rate precision of the prediction is separately around 89.47% and 97.91%and error rate is about 10.53% and 2.09%on average with the regression predication for static test. Therefore, the regression method is an acceptable method for the prediction of the pitting corrosion rate, and the prediction method should be explored besides other statistics analysis method for the pitting corrosion depth.

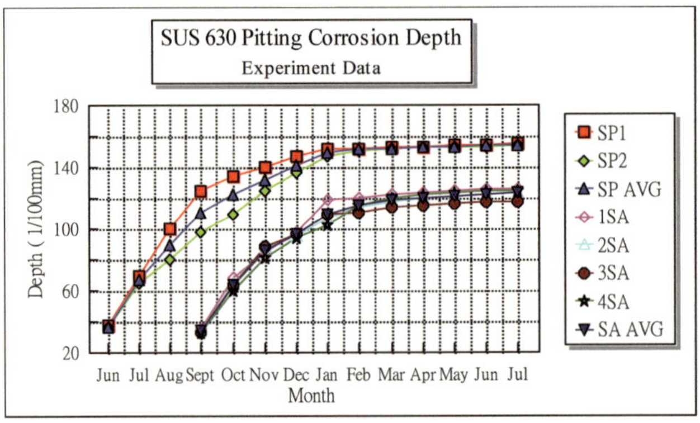

Fig. 33. Pitting corrosion depth of dynamic experiment and static experiment

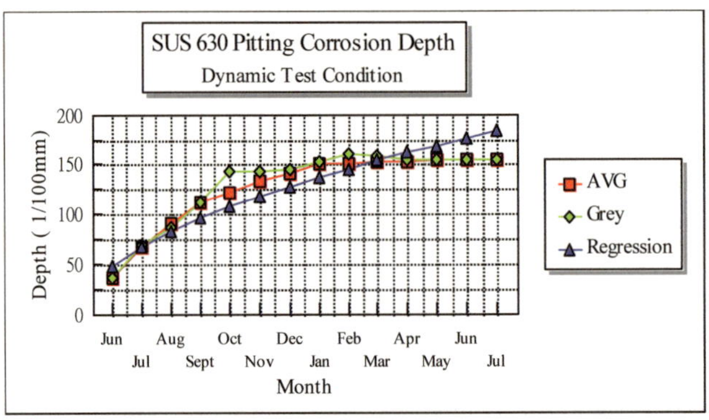

Fig. 34. Pitting corrosion depth of dynamic experiment with predict the curve graph by Grey and Regression

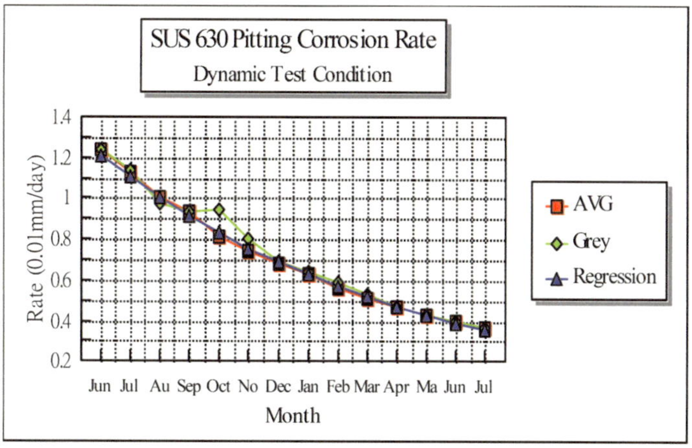

Fig. 35. Pitting corrosion depth of static experiment with predict the curve graph by Grey and Regression

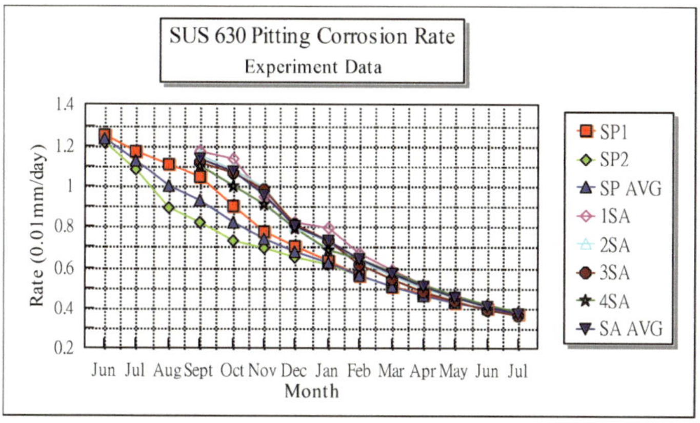

Fig. 36. Pitting corrosion ratio of dynamic experiment with predict the curve graph

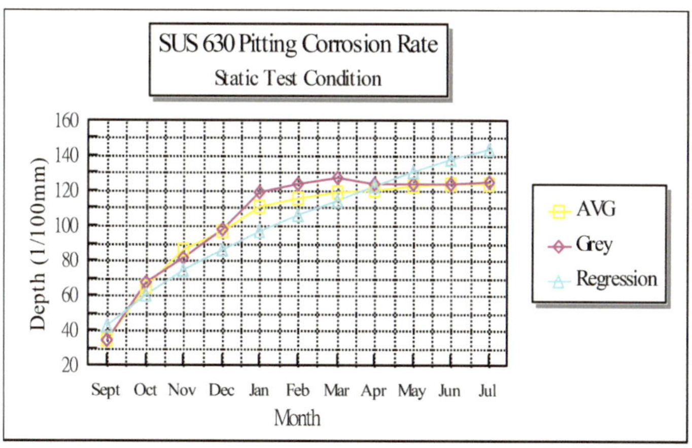

Fig. 37. Pitting corrosion rate of dynamic experiment with predict the curve graph by Grey and Regression

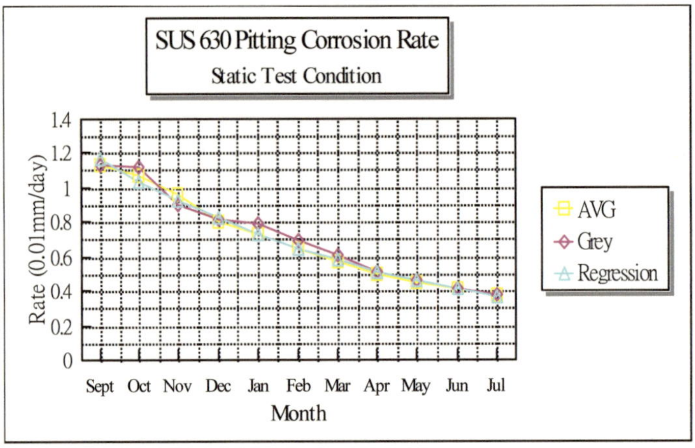

Fig. 38. Pitting corrosion rate of static experiment with predict the curve graph by Grey and Regression

5. Pitting corrosion fatigue life

As the condition of most pitting corrosion occurring on the surface of stainless steel shafts, when they are operating in fluid of neutral-to-acid solution, which especially containing chloride or chloride ions, there constitute multiple points of stress concentration on the shaft surface . The mechanism of multiorigin fatigue fracture may be more complicated than that of the above mentioned single point stress concentration condition.

5.1 Fatigue

In material science fatigue is the progressive and localized structural damage that occurs when a material is subjected to cyclic loading. The nominal maximum stress values are less than the ultimate tensile stress limit, and may be below the yield stress limit of the material.

A fatigue crack often starts at some point of stress concentration. This point of origin of the failure can be seen on the failed material as a smooth, flat, semicircular or elliptical region, often referred to as the nucleus. Surrounding the nucleus is a burnished zone with ribbed markings. This smooth zone is produced by the crack propagation relatively slowly through the material and the resulting fractured surfaces rubbing together during the alternating stressing of the component [1]. When the component has become so weakened by the crack that it is no longer able to carry the load, the final, abrupt fracture occurs, which shows a typically crystalline appearance.

Fatigue occurs when a material is subjected to be repeated loading and unloading. If the loads are above a certain threshold, microscopic cracks will begin to form at the surface. Eventually a crack will reach a critical size, and the structure will suddenly fracture. The shape of the structure will significantly affect the fatigue life; square deeps or sharp corners will lead to elevated local stresses where fatigue cracks can initiate. Round deeps and smooth transitions or fillets are therefore important to increase the fatigue strength of the structure.

According to the observation ways of a metal fracture, the fracture can be categorized into macroscopic and microscopic features. Marcroscopic fracture can be observed or inspected by either naked eyes or a magnifying glass on the surface of the metal or the fracture surface. While the microscopic fracture can be observed only by a optical microscope or a scanning electron microscope (SEM). Owing to the enlarging area and depth is not so extensive by a optical microscope, therefore a SEM is used in this study. On the contrary, SEM can observe a broad range and deeper depth on the fracture surface. Through microscopic fracture surface observation, the crack propagation process and material intrinsic crystalline structure can be identified.

5.2 Fatigue life

The fatigue life time depends on several factors, where the most important one sere the manufacturing, the material properties, and the loading conditions, which are all more or less random. Both material properties and dynamical load process are important for fatigue evaluation, and should in more realistic cases be modeled as random phenomena.

Since the well-known work of Wöhler in Germany starting in the 1850's, engineers have employed curves of stress versus cycles to fatigue failure, which are often called S-N curves (stress-number of cycles) or Wöhler's curve.The basis of the stress-life method is the Wöhler S-N curve, that is a plot of alternating stress, S, versus cycles to failure, N. The data which results from these tests can be plotted on a curve of stress versus number of cycles to failure. This curve shows the scatter of the data taken for this simplest of fatigue tests.

ASTM defines fatigue life (N), as the number of stress cycles of a specified character that a specimen sustains before failure of a specified nature occurs. Fatigue life is the number of loading cycles of a specified character that a given specimen sustains before failure of a specified nature occurs. When analyzing the fatigue life for the structures, the level crossings have been used for a long time. However, better life predictions are obtained when using a cycle counting method, which is a rule for pairing local minima and maxima to equivalent load cycles. Fatigue damage is computed by damage accumulation hypothesis which is illustrated as Palmgren-Miner rule. This rule is used to obtain an estimate of the structural fatigue life.

In high-cycle fatigue situations, materials performance is commonly characterized by an S-N curve, also known as a wdeep curve. This is a graph of the magnitude of a cyclic stress (S) against the logarithmic scale of cycles to failure (N).

S-N curves are derived from tests on samples of the material to be characterized where a regular sinusoidal stress is applied by a testing machine which also counts the number of cycles to failure. This process is sometimes known as coupon testing. Each coupon test generates a point on the plot though in some cases there is a run out where the time to failure exceeds that available for the test. Analysis of fatigue data requires techniques from statistic, especially survival analysis and linear regression.

5.3 S-N curve with rotating bending test

5.3.1 Fatigue rotating bending test

For this rotating bending testing method followed JIS Z 2247-1978, the material is stainless SUS 630 and the test pieces standard is symbol 1-10 of class 1, and the speed of loading repetition is 2400 rpm. The value obtained by diving the bending moment at a cross section generating the maximum stress by modulus of section shall be used as the nominal stress. The S-N curve will be drawn by plotting the values of stress amplitude as ordinate and the number of stress cycles. Configurations of the test machine and the specimen used in the study are shown as Figure 39 and Figure 40 respectively.

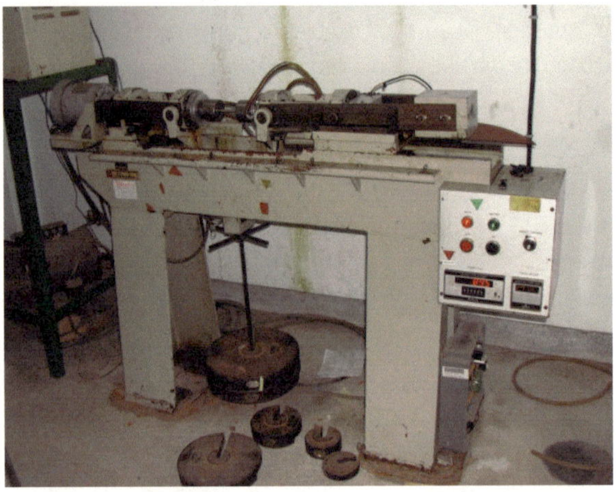

Fig. 39. Rotating bending fatigue test machine

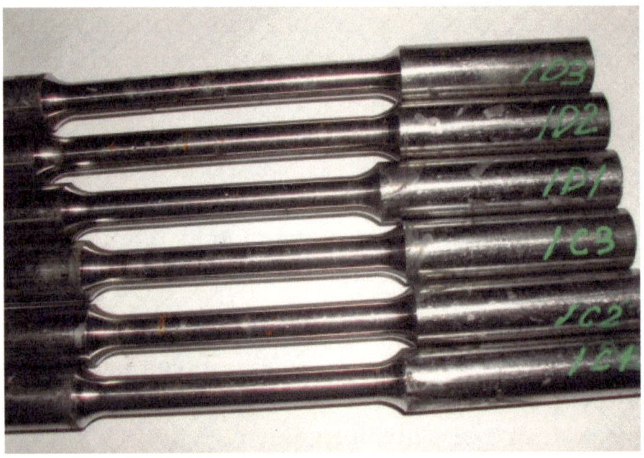

Fig. 40. Specimens for fatigue test under rotating bending load

5.3.2 Fatigue life estimation

In this test, the number of cyclic stress reversals with constant amplitudes of 350 N/mm^2, 400 N/mm^2, 450 N/mm^2, 500 N/mm^2, 550 $N/mm^2$and 600 N/mm^2, respectively, induced in the specimens is counted. In addition, the test for that on uncorroded specimens is also carried out. Hanging equivalent weight on the test specimen and rotating under the revolution speed of 2400 rpm attain these specified stress amplitudes.

Due to the yield stress around 725 $N/mm^2$and maximum stress amplitude 600 N/mm^2 in this fatigue test, therefore, the fatigue life should be estimated with statistic method for over 600 N/mm^2 stress. According to the eempirical rule, the estimated theory will be selected the Grey Model GM (1,1) and Curvilinear Regression same as Section 4.

Using the experimentation data as input, a computer program was used to obtain the results in Tables 5.1. The S-N curve showing the experimentation, grey predication and regress analysis for SUS 630 fatigue test sample are shown in Figure 41 to Figure 42.

Stress (N/mm²)	Experiment	Grey Predicate	Regress Analysis
750		26503	12417
700		42784	24998
650		70640	50327
600	153500	119415	101320
550	218500	206914	203984
500	247800	367913	410670
450	523300	672130	826782
400	1800000	1263187	1664520
350	5000000	5000000	3351095

Table 8. Experiment, Grey Predication and Regress Analysis value

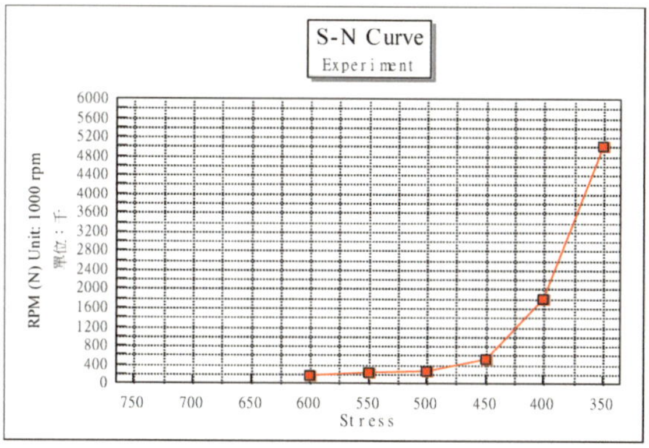

Fig. 41. S-N Curve with experiment

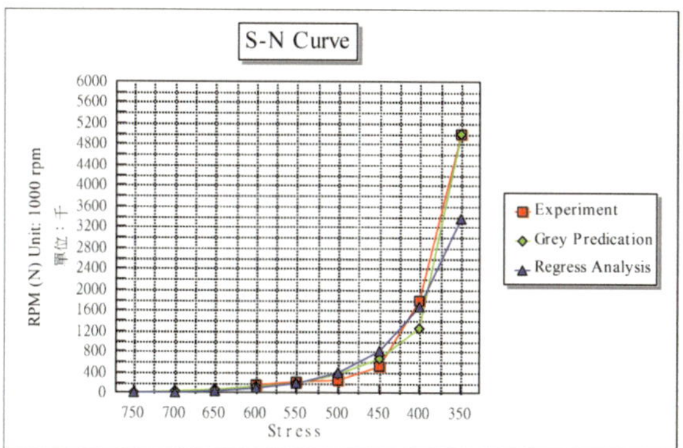

Fig. 42. S-N Curve with Experiment, Grey Predication and Regress Analysis

The precision of the prediction is around 98.31875 with the grey predication;, the precision of the prediction is around 97.54732 with the regress analysis. Therefore, the grey model GM (1,1) is an acceptable method for the prediction of the S-N curve.

The grey system theory has been used to deal with partially known and unknown data for stainless alloy. Traditional prediction usually employs the statistical method for large samples of data. According to the above-mentioned results, the grey system theory is also suitable for use to predict the S-N curve the scarcity of available data.

5.4 S-N curve under pitting corrosion condition

5.4.1 Ferric Chloride Test (FCT)

In order to accelerate the generation of SUS 630 shaft specimen with various extents pitting corrosion to facilitate the fatigue crack propagation study, the ferric chloride test (FCT) in accordance with the standard of JIS G 0578: 2000, is performed. Besides, the ferric chloride test can be used to evaluate the pitting resistance of stainless steels. At first, the test specimens were put in the ferric chloride solution with concentration 6% and carry out continuous immersion for 24hours, 48hours and 72hours respectively. Thus, three sets of specimen are prepared. They are specimen No. 3A1, 3A2, and 3A3 attained from 24h FCT; specimen No. 1A1, 1A2, and 1A3 from 48h FCT; and specimen No. 2A1, 2A2, and 2A3 from 24h FCT.

After the 24h, 48h and 72h immersion, the test specimens are taken out from the test solution. After removing the crusts generated by corrosion and a dial gauge measures washing and drying the specimens, the pitting depth at five representative points. The measured depths together with the average pitting depth are shown in Table 9.

	24 h FCT	48 h FCT	72h FCT
Specimen No.	3A1	1A1	2A1
Average depth (mm)	0.31	0.57	0.71
Specimen No.	3A2	1A2	2A2
Average depth (mm)	0.44	0.59	0.72
Specimen No.	3A3	1A3	2A3
Average depth (mm)	0.45	0.66	0.80

Table 9. Measured record and average pitting depth of test specimen

5.4.2 Fatigue life of pitting corrosion

The fatigue test was carried out with rotating bending test equipment under the number of cyclic stress reversals with constant amplitudes of 350 N/mm^2, 400 N/mm^2 and 450 N/mm^2, respectively, induced in the specimens is counted.

The fatigue life (N_f) can be divided into the number of cycles, (N_i) necessary to initiate a stage I micro crack, which has a size of the order of the grain size, and the number of cycles, (N_p) for crack propagation. Then a crack growth law relating the propagation rate of each microcrack to the applied plastic strain and crack length is derived metallographic observations. The third stage of fatigue damage, which leads to final fracture, is modeled as the coalescence of this population of microcracks which are nucleated and which propagate continuously. The measurement data for the fatigue test belong to the N_p in Table 10 to Table 13.

24 Hours		
Specimen No.	Test stress range (N/mm²)	No. of cycles(cycles)
3A2	450	144200
3A1	400	155200
3A3	350	244100
48 Hours		
Specimen No.	Test stress range (N/mm²)	No. of cycles(cycles)
1A2	450	103200
1A1	400	124100
1A3	350	151200
72 Hours		
Specimen No.	Test stress range (N/mm²)	No. of cycles (cycles)
2A2	450	64000
2A1	400	97600
2A3	350	123500

Table 10. Measurement data of the experiment (without any treatment)

24 Hours		
Specimen No.	Test stress range (N/mm²)	No. of cycles(cycles)
2B2	450	14700
2B1	400	32700
2B3	350	47800
48 Hours		
Specimen No.	Test stress range (N/mm²)	No. of cycles(cycles)
1B2	450	11700
1B1	400	18500
1B3	350	33200
72 Hours		
Specimen No.	Test stress range (N/mm²)	No. of cycles (cycles)
3B1	450	
3B2	400	20300
3B3	350	26900

Table 11. Measurement data of the experiment (with solution treatment)

24 Hours (coating surface with machined)		
Specimen No.	Test stress range (N/mm²)	No. of cycles(cycles)
2C2	450	14500
2C1	400	27400
2C3	350	63700
48 Hours (coating surface with machined)		
Specimen No.	Test stress range (N/mm²)	No. of cycles(cycles)
1C2	450	14100
1C1	400	27100
1C3	350	62700
72 Hours (coating surface without machined)		

Specimen No.	Test stress range (N/mm²)	No. of cycles (cycles)
3C2	450	78700
3C1	400	94300
3C3	350	142100

Table 12. Measurement data of the experiment (with metal coating treatment)

24 Hours (coating surface with machined)		
Specimen No.	Test stress range (N/mm²)	No. of cycles(cycles)
2D2	450	21600
2D1	400	38500
2D3	350	65900
48 Hours (coating surface with machined)		
Specimen No.	Test stress range (N/mm²)	No. of cycles(cycles)
1D2	450	13500
1D1	400	25100
1D3	350	27600
72 Hours (coating surface without machined)		
Specimen No.	Test stress range (N/mm²)	No. of cycles (cycles)
3D2	450	105400
3D1	400	134800
3D3	350	670000

Table 13. Measurement data of the experiment (with metal coating treatment)

5.4.3 Fatigue behavior of pitting corrosion

The test results of the number of cyclic stress reversals till fatigue fracture of the specimens are listed in Table 14. In Table 14, the fatigue lifetime ratio of pitting corrosion condition of the specimen is only about 12 - 27% of the fatigue time of uncorroded condition for the case of the working stress variation range 450 N/mm². the range of fatigue lifetime ratios even lower to 5.2 - 8.6 % and 2.5 - 4.8% for stress variation amplitudes 400 N/mm² and 350 N/mm² stress amplitude respectively.

Stress amplitude (N/mm²)	No. of cyclic stress reversals of specimen				Fatigue life ratio
	Uncorroded	24h FCT	48h FCT	72h FCT	
600	153500				
550	218500				
500	247800				
450	523300	144200	103200	64000	12 - 27 %
400	1800000	155200	124100	97600	5.2 - 8.6%
350	5000000	244100	151200	123500	2.5 - 4.8%

Table 14. Fatigue life test on specimens of uncorrod condition and pitting corrosion condition

6. Material constant

Failures occurring under condition of dynamic loading are called fatigue failures, presumably because up to the present it has been believed that these failures occur only after a considerable period of service. Fatigue is said to account for at least 90 percent of all service failure due to a mechanical cause.

Fatigue results in a brittle-appearing fracture, with no gross deformation at the fracture. On a macroscopic the fracture surface is usually in line with the direction of the principal tensile stress.

A fatigue failure can usually be recognized from the appearance of the fracture surface, which shows a smooth region, due to the rubbing action caused by the crack propagation through the section; and a rough region, due to the member failing in a ductile manner caused by the cross section being no longer able to carry the load.

6.1 Fatigue crack propagation

Considerable research has gone into determining the laws of fatigue crack propagation for stage II growth. The crack propagation rate da/dN is found to follow an equation:

$$\frac{da}{dN} = C \cdot \sigma_a^m \cdot a^n \tag{1}$$

where
C: constant
σ_a: alternating stress
a: crack length

The most important advance in placing fatigue crack propagation into a useful engineering context is the realization of the fact that crack length versus cycles at a series of different stress levels could be expressed by a general plot of da/dN versus ΔK. da/dN is the slop of the crack growth curve at a given value of a, and ΔK is the range of the stress intensity factor, defined as

$$\begin{gathered} \Delta K = K_{\max} - K_{\min} \\ \Delta K = \sigma_{\max}\sqrt{\pi a} - \sigma_{\min}\sqrt{\pi a} = \sigma_r\sqrt{\pi a} \end{gathered} \tag{2}$$

Since the stress intensity factor is undefined in compressions, K_{min} is taken as zero if σ_{min} is compression.

Region II represents an essentially linear relationship between log da/aN and log ΔK

$$\frac{da}{dN} = C(\Delta K)^n \tag{3}$$

where:

$$\Delta K = \Delta\sigma(\pi a)^{\frac{1}{2}} Y \qquad \Delta K = A(a)^{\frac{1}{2}}$$

For this empirical relationship n is the slop of the curve and A is the value found by extending the straight line to $\Delta K = 1$ MPa $m^{1/2}$.

The equation (3) is often referred to as Paris's law. The equation provides an important link between fracture mechanics and fatigue.

Estimation the Fatigue crack propagation

If let the fatigue crack propagation for each cycle revolution is μ, then,

$$\mu = \frac{da}{dN} \tag{4}$$

or

$$dN = \frac{da}{\mu} \tag{5}$$

where:

a crack depth

N cycle revolution

First, measure the pitting deep depth and assume the depth is the crack beginning depth (a_0), then, measure the depth of the smooth area and assume the depth is the fatigue propagation depth (a_p), next, get the beginning depth plus the fatigue propagation depth and assume the depth is the critical depth (a_c). Finally, use the equation (6) and find the life cycle of the fatigue propagation (N_p).

$$N_p = \int_{a_0}^{a_c} \frac{da}{\mu} \tag{6}$$

where:

a_0 the depth of crack beginning propagation

a_c the critical depth

The crack propagation ratio is show in the formula (7) when the structure has been loaded with a stable alternating stress.

$$\begin{aligned} \frac{da}{dN} &= C_1 \Delta K^{n_1} \qquad 10^{-6} < \frac{da}{dN} < 10^{-4} \qquad (\mathrm{mm/N}) \\ \frac{da}{dN} &= C_2 \left(\Delta K - \Delta K_{th}\right)^{n_2} \qquad 0 < \frac{da}{dN} < 10^{-6} \qquad (\mathrm{mm/N}) \\ \frac{da}{dN} &= 0 \qquad \Delta K < \Delta K_{th} \end{aligned} \tag{7}$$

where:

C_i, n_i (i = 1,2…) material constant coefficient

The Paris formula can be deduced from the relations between the ratio of fatigue propagation and the crack depth as follows:

$$\frac{da}{dN} = C(\Delta K)^n \tag{8}$$

$$\Delta K = \Delta\sigma(\pi a)^{\frac{1}{2}} Y \tag{9}$$

where:

C, n	material constant;
Y	geometry factor;
$\Delta\sigma$	the difference between maximum stress (σ_{max}) and minimum stress (σ_{min});
a	crack depth

with the giving structure and the constant alternate load ($\Delta\sigma$), the equation (9) can be rewritten as follows:

$$\Delta K = A(a)^{\frac{1}{2}} \tag{10}$$

where:

$A = Y \cdot \pi^{\frac{1}{2}} \Delta\sigma = const$

$$\mu = \frac{da}{dN} = C(A\sqrt{a})^n = C_0 a^{\frac{n}{2}} \tag{11}$$

where:

$C_0 = CA^n$

$$N_p = \int_{a_0}^{a_c} \frac{da}{C_0 a^{n/2}} = \frac{2}{(2-n)C_0} a^{1-\frac{n}{2}} |_{a_0}^{a_c} = \frac{2}{(2-n)C_0} [a_c^{1-\frac{n}{2}} - a_0^{1-\frac{n}{2}}] \tag{12}$$

The constant C_0 and n can be decided with the following method, if, both sides of the equation (11) are taken with log function as follows:

$$\log\left(\frac{da}{dN}\right) = \log C_0 + (\frac{n}{2})\log a \tag{13}$$

so,

The relation between $\log\left(\frac{da}{dN}\right)$ and $\log a$ is linear, the intercept is $\log C_0$; the slop is $\frac{n}{2}$,

Therefore, the corresponding value is $\left(\frac{da}{dN}\right)_i$ for the different crack depth (a_i), it varies according to the equation (13) and act as a go-between to all or division. The captioned formula can be derived as C_0 and $\frac{n}{2}$, and will be able to comply with equation (12), then N_p can be derived from equation (12).

6.2 Methodology

The Ferric Chloride Test (FCT) has been carried out same as the above 5.3.1 for the pitting corrosion and the fatigue tests are carried out for the sampling pieces; then, the metallurgical graphics are taken through the Scanning Electron Microscope (SEM) for the fatigue fracture surface.

The fatigue test result data and measuring the initial crack and critical crack depth from the SEM graphic are shown in the Table 15, Table 16 and Table 17.

Specimen No.	Test stress	Fatigue revolution (rpm)	Initial crack depth (a_0)(mm)	Critical crack depth (a_c)(mm)
3A2	450	144200	0.78	4.15
3A1	400	155200	1.23	3.26
3A3	350	244100	1.07	4.85

Table 15. 24h FCT fatigue test data and measuring record

Specimen No.	Test stress	Fatigue revolution (rpm)	Initial crack depth (a_0)(mm)	Critical crack depth (a_c)(mm)
1A2	450	103200	1.20	4.47
1A1	400	124100	1.13	3.83
1A3	350	151200	1.53	4.13

Table 16. 48h FCT fatigue test data and measuring

Specimen No.	Test stress	Fatigue revolution (rpm)	Initial crack depth (a_0)(mm)	Critical crack depth (a_c)(mm)
2A2	450	64000	1.10	3.77
2A1	400	97600	0.70	2
2A3	350	123500	1.00	4.52

Table 17. 72h FCT fatigue test data and measuring record

6.3 Material constant

Figure 43 show the metallurgical graphics of the fatigue fracture surface with the SEM. The graphics allow to measure the initial crack depth (pitting deep depth) and to calculate the critical crack depth. The result is shown as Table 15 to Table 17.

	450 N/mm^2	400 N/mm^2	350 N/mm^2
A	797.60	708.98	620.36
K	704.425	786.30	620.36
C0	1.17E-6	1.08E-6	1.05E-6
C	1.72E-14	3.47E-15	4.40E-15
n	2.7	2.98	3.0
da/dN	8.39E-7	1.47E-6	1.16E-6

Table 18. SUS630 Material Constant and Fatigue Crack Propagation Constant

The fatigue crack propagation can be estimated using the data of Table 15 to Table 17 and the results that can be obtained as shown in Table 18. Table 18 gives the ranges of constant n

(2.7 to 3.0), constant (C 3.47E-15 to 1.72E-14) and da/dN (8.39E-7 to 1.47E-6). The tendency curves of the constant n, C and da/dN are shown in Figure 44 to Figure 46.

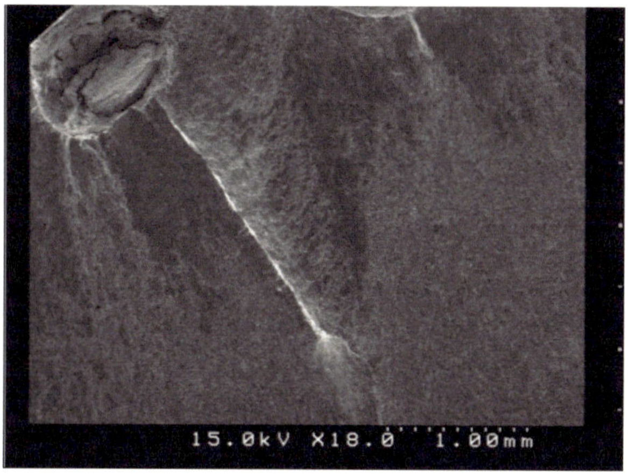

Fig. 43. Fatigue propagation area

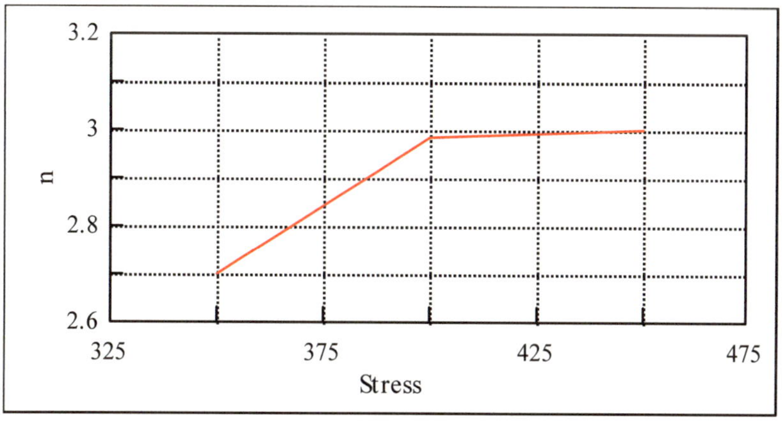

Fig. 44. n Tendency Curve

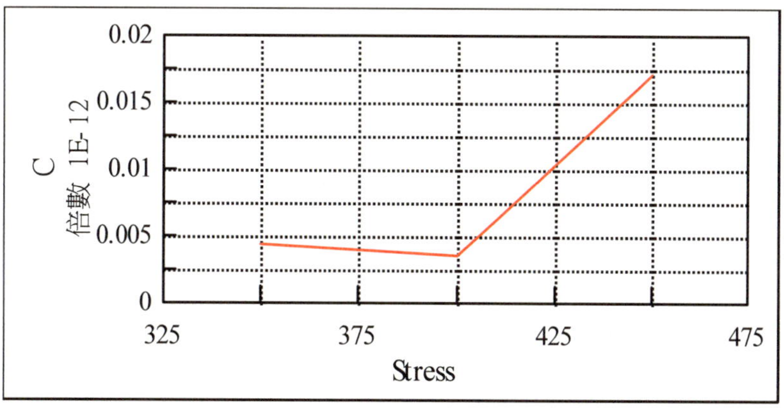

Fig. 45. C Tendency Curve

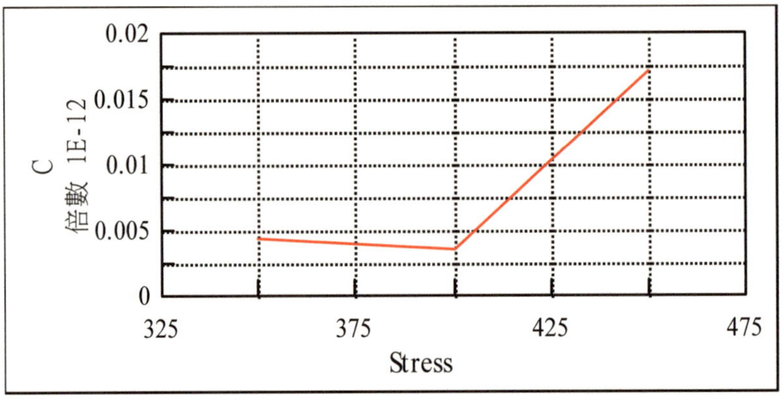

Fig. 46. da/dN Tendency Curve

7. Summery

Present knowledge about the relations between the microstructure changes and durability is still limited. This is the complex problem and it is necessary to deal with it as the separate question in the low-cycle fatigue durability estimation methodology. Material properties change in the low-cycle fatigue process, because most of the structural materials as steel for example are metallurgical unstable when they are cyclically deformed particularly at high temperature. Their dislocation structure changes determine the crack initiation in steel and their macroscopic behavior.

The fatigue lifetime of SUS 630 shaft under various extent of pitting corrosion condition is found to be in a range of only 2.5~27% of that of the uncorroded condition. These results are obtained by the ferric chloride test and rotating bending fatigue test on the SUS 630 shaft specimen. Apart from the feature of common fatigue fracture starting from a single point of stress concentration, i.e., forming initial stress nucleus, propagating ribbed markings and complete fracture zone, there existed much more sophisticated features on the fatigue fracture surfaces generated by multi points of stress concentration around the pitting deeps. Such as the fatigue lines, the fatigue beach marks, the fatigue stair lines, the critical crack length, the river stream pattern, tongue pattern and tree pattern in fracture area, transgranular and intergramular fractures, etc., have been detected and identified by using a SEM in the study.

Fatigue life showed difference depending on the matrix structure. The apparent difference of the fatigue life between the stress and the strain controlled cycling was well correlated in each material by means of cyclic stress-strain relationship. The grey system theory has been used to deal with partially known and unknown data for stainless alloy. Traditional prediction usually employs the statistical method for large samples of data. According to the above-mentioned results, the grey system theory is also suitable for use to predict the S-N curve the scarcity of available data.

Any of the three basic factors may cause fatigue failure list below. These are (1) a maximum tensile stress of sufficiently high value, (2) a large enough variation or fluctuation in the applied stress, and (3) a sufficiently large number of cycles of the applied stress. In addition, there are a host of other variables, such as stress concentration, corrosion, temperature, overload, metallurgical structure, residual stresses, and combined stresses, which tend to alter the conditions of fatigue. Since we have not yet gained a complete understanding of what causes fatigue in metals, it will be necessary to discuss each of these factors from an essentially empirical standpoint. Due to the availability of a great mass of data, it would only to describe the highlights of the relationship between these factors and fatigue.

Constant n will be increased when the stress strength is increased. The rise ratio is rapid when the stress strength is less than the fatigue stress, the rise ratio is slow when the stress strength is bigger than the fatigue stress; Constant C will go down when the stress strength is less than the fatigue stress, constant C will go up when the stress strength is bigger than the fatigue stress; Constant da/dN will go up when the stress strength is less than the fatigue stress, The constant da/dN will be decreased when the stress strength is bigger than the fatigue stress. The fatigue strength is a breakpoint for the material constants n, C and da/dN of the fatigue fracture for the stainless SUS 630.

8. References

[1] Chin-Ming Hong Lon-Biou Li A Study of the Application of the Grey Correlation and the Analysis of Excellence Journal of Technology, Vol. 12 No. 1 1997 (in Chinese)

[2] Chang-Huang Chen A New Form of Grey Relational Coefficient Journal of Tung Nan Institute of Technology 2002

[3] Denny A. Jones Principles and Prevention of Corrosion Prentice Hall Inc. (1996)

[4] Der Norske Veritas Annual Report The New Risk Reality 2002

[5] Fong Yuan Ma ,Wei Hui. Wang A Study of the Relation between Economic Growth Rate and Shipping Industry Growth Rate Prediction and based on the Grey System Theory The conference of Cross-Strait for Navigating and Logistic 2004

[6] Fong Yuan Ma A Study of Tendency of Typhoon Landing in Taiwan of Using GM(1,1) Model The conference of Cross-Strait for Seamen 2004.

[7] Fong Yuan Ma and Wei Hui Wang, Fatigue crack propagation estimation of SUS 630 shaft based on fracture surface analysis under pitting corrosion condition, Material Science & Engineering A430, 2006, 1-8.

[8] Fong Yuan Ma and Wei Hui Wang, Prediction of pitting corrosion behavior for stainless SUS 630 based on grey system theory, Material Letters, 2006.

[9] Fong Yuan Ma ,Wei Hui. Wang *Estimating the Fatigue Life by Grey Theory for Propulsion Shaft of High Speed Craft* The 7th International Symposium on Marine Engineering conference 2005

[10] Fong Yuan Ma *Design Criterion Evolution of Shipboard Propulsion Shafting System Based on Classification Rules* The TEAM Conference 2005

[11] George E. Dieter, Mechanical Metallurgy, McGraw-Hill Book Company, 1988.

[12] G.S. Frankel, Pitting Corrosion Metals Handbook, Ohio State University 1994

[13] Japanese Standards Association, JIS Handbook, Ferrous Materials and Metallurgy I, 2002.

[14] J. L. Deng Basic of Grey System Theory Huazhong University Science and Technology Press, 2003 (in Chinese)

[15] Joseph F. Healey *Statistics: A tool for Social Research* Wadsworth Publishing Company 1995

[16] Jyh-Horng Wen and Yung-Fa Huang Application of Grey Prediction on Cellular Mobile Communication Systems IICM Vol. 3 No. 1 2000

[17] K. Enami, Y. Hagiwara, H. Mimura Assessment Method of Ductile and Brittle Fracture Initiation in High Strength Steels Journal of the Society of Naval Architects of Japan Vol. 195 2004

[18] M. Mohri & others A System Development in Predicating Fatigue Crack Paths and its Verification by Fatigue Tests Journal of the Society of Naval Architects of Japan Vol. 194 2003

[19] Petinov, S., Fatigue Analysis of Ship Structures, Backborne Publishing Company, 2003.

[20] P. Shi and S. Mahadevan Corrosion fatigue and multiple site damage reliability analysis International Journal of Fatigue June 2003.

[21] Ruey-Hsun Liang A Hybrid Grey System-Dynamic Programming Approach for Thermal Generating Unit Commitment Journal of Science and Technology Vol. 8 No. 1 1999

[22] SECIL ARIDURU Fatigue life calculation by rainflow cycle counting method 2004.

[23] Thomas Leonard, John S. J. Hsu *Bayesian Methods: An Analysis for Statisticians and Interdisciplinary Researcher* Cambridge University Press in 1999

[24] Walter A. Rosenkrantz *Introduction to Probability and Statistics for Scientists and Engineers* The Mcgraw-Hill Companies, Inc. 1995

[25] William Bolton, Engineering Materials 2, Newness ed, 1993